NEXUS NETWORK JOURNAL

MECHANICS IN ARCHITECTURE

In memory of Mario Salvadori 1907-1997

VOLUME 9, NUMBER 2

Autumn 2007

KIM WILLIAMS BOOKS

Nexus Network Journal
Vol. 9
No. 2
Pp. 159-382
ISSN 1590-5896

CONTENTS

*M*ario Salvadori in many ways embodied architecture and mathematics, and certainly embodied the particular mental aperture necessary for interdisciplinarianism. I first met Mario in 1991, not long after I had moved to Tuscany. Like all architects in my generation, I had studied structural mechanics from his textbooks at university. When I moved to the province of Florence, I discovered that Salvadori was a rather common name; I wondered if it might be the same family. Once when I returned to New York I called him, though of course he didn't know who I was, and explained where I was from. He invited me to come to his office that very day and have a sandwich in his office. That was how our friendship began, and it is characteristic of the kind of spontaneity, warmth and interest that Mario always exuded. Already in his 80s when we met, he was still as bright as a dollar, continuing his writing and teaching at the Salvadori Center. He encouraged me every step of the way as I organized the first Nexus conference for architecture and mathematics in 1996. He came, with his wife Carol, to that conference, in Fucecchio, not far from Legnaia where he was born, and gave the keynote address, entitled "Are There Any Relationships Between Architecture and Mathematics". The next year he passed away. I have wished many times he could have seen how the Nexus conferences grew, and how the *Nexus Network Journal* was founded and prospered.

This year is both the one-hundredth anniversary of Mario's birth and the tenth anniversary of his death, and we are pleased to honor him with this issue. Pietro Nastasi and I tell part of his story in "Mario Salvadori and Mauro Picone: From Student and Teacher to Professional Fellowship."

The Nexus 2006 conference in Genoa, Italy, included a special session dedicated to "Mechanics and Architecture". This was specially conceived to honor the Genovese scholar who did so much to advance historical studies in the science of construction, Edoardo Benvenuto. Former students and colleagues of Benvenuto, who passed away in 1998 founded the "Associazione Edoardo Benvenuto for the research on the Science and Art of Building in their historical development" in Genoa, which carries on Benvenuto's passion for and particular point of view of historical studies of relationships between mechanics and architecture. Four of the papers in this issue of the *Nexus Network Journal* dedicated to this theme were presented at the Genoa conference.

In the panorama of relationships between architecture and mathematics, structural mechanics occupies a special place, though not a universally appreciated one. Of the architects and lovers of architecture who recognize the value of mathematics as a powerful formal tool to generate shape, promote coherent systems of proportions, define relationships of distinct parts through symmetries, and embue architecture with meaning, some will argue that mechanics – including the study of equilibrium and the behavior and computation of the internal forces in a structure – belongs not to architecture but rather to engineering. This is certainly not a new point of view, but one which is not consonant with an interdisciplinary approach. Architects and engineers sometimes collaborate, but at other times view each other with suspicion, each considering the other as a necessary evil (it amused me recently to read in Mario Salvadori's *Mathematics in Architecture*, where he advises the architect to learn how to use the slide rule, "By learning how to perform elementary operations on the slide rule, you will become much more confident in your work and different person in the eye of the engineer" [Prentice-Hall, 1968, p. 143]). But almost always the line of demarcation between the two fields is clear: to the architect fall the formal and aesthetic considerations; to the engineer the technical considerations.

Nexus Network Journal 9 (2007) 163-164   NEXUS NETWORK JOURNAL – VOL. 9, No. 2, 2007 **163**
1590-5896/07/020163-2 DOI 10.1007/s00004-006-0037-y

However, the papers presented here show how interrelated the aesthetic and the technical considerations are. Marco Giorgio Bevilacqua, in "Ramparts in the Sixteenth century: Architecture, 'Mathematics' and Urban Design", examines how developments in techniquees of warfare, including the discovery and widespread use of gunpowder, rendered earlier forms of defensive architecture obsolete, and in particular how the shape in plan of fortresses changed in response to the need to deflect assault from new weapons. In "*Tentare licet.* The Theresian Academy Question on the Theory of Beams of 1783", Dirk Van de Vijver takes us back to a key moment in history, a kind of turning point, when the methods of calculating internal forces in structural elements were just being discovered and formalized, and when such endeavors were looked on with scant enthusiasm by those in the building trades. Santiago Huerta examines the history, geometry and mechanics of "Oval Domes", ancient forms that became the leitmotif of Baroque architecture. Federico Foce presents "Milankovitch's *Theorie der Druckkurven*: Good Mechanics for Masonry Architecture", an almost poignant example of a careful theoretical analysis that in a sense arrived too late, that is, after the eclipse of masonry by new materials such as reinforced concrete and steel. Olivier Baverel and Hoshyar Nooshin present a rigorous examination of "Nexorades Based on Regular Polyhedra". This is an example of a structural system that presents noteworthy aesthetic possibilities.

In other research in this issue of the *NNJ*, Frans Cerulus presents "A Pyramid Inspired by Mathematics", an eighteenth-century architectural "folly" with remarkable proportions and acoustic properties. Yannick Joye reexamines a theme already treated several times in our pages. In "Fractal architecture Could Be Good for You" he argues that human beings are psychologically and physiologically attuned to fractal systems. Antonia Redondo Buitrago has studied an irrational value to point out new relationships between "Polygons, Diagonals and the Bronze Mean". This could challenge architects to find new uses for the metallic means in design.

The subject of Rachel Fletcher's Geometer's Angle column in this issue is "Dynamic Root Rectangles." This is Part I, "The Fundamentals", to be followed by advanced properties and applications in the next issue.

In Didactics, Igor Verner and Sarah Maor present "Mathematical Aspects in an Architectural Design Course: The Concept, Design Assignments, and Follow-up", This is itself a follow-up to the discussion begun by the authors in the *NNJ* vol. 8 no. 1.

The issue concludes with two book reviews. Rachel Fletcher reviews Scott Olsen's *The Golden Section: Nature's Greatest Secret*. Scott Olsen is a past contributor to the *NNJ* (see vol. 4, no. 1). It was a great pleasure for me to review *Hardy Cross: American Engineer* by Leonard K. Eaton, timely because of the theme of this issue. Prof. Eaton first published a paper on Hardy Cross in the *NNJ* vol. 3 no. 2, and his book grew out of that research.

I hope this issue helps you discover unexpected aspects of the relationships between architecture and mathematics, even *engineering* mathematics!

Kim Williams

Kim Williams

Via Cavour, 8
10123 Turin (Torino) Italy
kwilliams@kimwilliamsbooks.com

Pietro Nastasi

Dipartimento di Matematica
Università di Palermo
Via Archirafi, 34
90123 Palermo ITALY
nastasi@math.unipa.it

Keywords: Mario Salvadori, Mauro
Picone, Istituto per le Applicazioni
del Calcolo, numerical analysis

Research

Mario Salvadori and Mauro Picone: From Student and Teacher to Professional Fellowship

Abstract. The correspondence between Mario Salvadori and Mauro Picone during the years 1934-1972 sheds light on the history of the Italian Institute for the Applications of Calculation. The IAC was a groundbreaking institution for mathematics in Italy, and great attention was given to the new means of mechanical calculation, first analogue, then electronic. It was in relationship to this that Mario Salvadori consulted with his former professor. The correspondence allows us to see also how that relationship changed from one of student-teacher to one of fellowship between professionals.

The books and individuals from whom I have learned what I know of numeric calculus are too numerous to list here, but I wish to express now my gratitude to Prof. Mauro Picone, director of the Istituto Nazionale per le Applicazioni del Calcolo (the Italian laboratory for applied mathematics which will house the International Center of Mechanical Calculus, recently founded), who was the first to teach me to love numbers while I was his student at the University of Rome some 20 years ago.

M. Salvadori, Preface to *Numerical Methods in Engineering*,
New York, Prentice-Hall, 1952

Introduction

This year, 2007, marks both the hundredth anniversary of the birth and the tenth anniversary of the death of Mario Salvadori (1907-1997), engineer, educator, and mathematician. It is no exaggeration to say that Salvadori's textbooks on structural engineering formed an entire generation of architects in the United States, and since their translation into other languages, in other nations as well. It also marks an important anniversary for another mathematician, Mauro Picone (1885-1977), who eighty years ago this year founded in Naples the embryo of a group for mathematics that would grow to be the Istituto per le Applicazioni del Calcolo "Mauro Picone" (*IAC*, the Italian Institute for the Applications of Calculation). It was the first such institute for calculation in the world. In Rome, Viale del Politecnico 137, near the site of the *Breccia di Porta Pia* (where on 20 September 1870 the State of the Catholic Church was defeated) is the present home of the *IAC*, founded in Naples in 1927 and transferred to Rome in 1932 under the aegis of Italy's CNR, the National Research Council, when Picone assumed a chair in the Department of Mathematics at the University of Rome.

A fortuitous (and fortunate) circumstance – that is, the accidental discovery of the basements of the Roman headquarters of the *IAC*, and in those basements, the rich documentation of past undertakings of the Institute – has allowed us to shed new light on the activities of Mauro Picone and his collaborators, including Mario Salvadori, who played a major role. The historic archives of the *IAC*, recently brought to light by the present director, Prof. Michiel Bertsch and edited by Maurizio Mattaliano, includes – among other documents – many letters exchanged by Picone and Salvadori between the years 1934 and 1972. This is one fundamental source for the present paper, which will trace the relationship between the two men. The other fundamental source is Salvadori's unpublished autobiography, entitled *A Tangential Life*.[1]

Mauro Picone and Mario Salvadori knew each other well, for Picone had been first Salvadori's professor and then thesis advisor at the University of Rome. In 1934, the year after Salvadori earned his doctorate in mathematics, he assumed a part-time position at the *IAC* under the direction of Picone.

This collaboration would turn out to be precious for both: Salvadori represented a window to the Anglo-Saxon scientific world due to his mastery of the language, and the *IAC* represented for Salvadori an excellent means of introduction into that world, where he quickly learned how to make his scientific merits appreciated. World events were interrupt Salvadori's career path, and take him from Italy to the United States. In spite of this, the professional lives of the two would remain close, and they remained close personally as well.

Mario Salvadori

Best known as a practicing engineer on the one hand and educator and author of textbooks of structural mechanics and engineering on the other, Salvadori was actually as much a mathematician as an engineer. Salvadori's early passions were music and mountain climbing, two activities frowned upon by his family. Mathematics was certainly not his first love; he writes, "Like the dishes one hates as a child and learns to like in later years, mathematics did not come naturally to me, but ended up by playing an important role in my life" [p. 3.1]. He describes four phases of his mathematical studies: the first, before he was seven years old; the second, when he was home-schooled by his father, Riccardo, an engineer; the third, his degree work in civil engineering; the fourth, his degree work in pure mathematics. Too intimidated to tell his father that he didn't grasp the concepts he presented to him, Salvadori writes, "I began first to fear and then to hate math" [p. 3.1]. Although he secretly wished to study music, dreaming of becoming an orchestra conductor, he bowed to pressure from his father and enrolled at the University of Rome in 1925 in engineering. The degree program was divided into two parts. The first two years were common to all specialties. His professors during this time there were of the highest caliber: he studied analytical geometry with Guido Castelnuovo (1864-1952); calculus with Francesco Severi (1879-1961); with Tullio Levi-Civita (1873-1941) he studied undergraduate rational mechanics. The last three years were directed toward particular fields; Salvadori's choice was civil engineering. As he describes the program:

> ...[T]he curriculum in civil engineering was both basic and wide: it emphasized structures and civil works (roads, canals, dams), but included fairly encompassing courses in electrical and mechanical engineering, and in

architectural design. We were thus better rounded technologists than the students of a contemporary American school [p. 16.21].

But he gives a rather laconic reason for choosing civil engineering: "[M]y mild interest in structures was so great in comparison with my complete lack of interest in all other subjects..." [p. 16.22]. He found the professors during the last three years to be uninspiring; among them were Anselmo Ciappi and Aristide Giannelli. He earned his degree in 1930 (the Italian *laurea* being equivalent to the doctorate in the United States). Of this first accomplishment he writes,

> In retrospect my successful graduation from the school of engineering was quite an achievement, because civil engineering at the school was 80% mathematics and I still did not really understand it. I could go mechanically through some mathematical proofs, knew some applied mathematics techniques, could even formulate some physical problems in mathematical terms, but the essence of the game still escaped me... [p. 3.4].

Announcing to his parents that he would *never* practice engineering, he decided to begin anew with a graduate course of study in pure mathematics. The reason will surprise many: "What started the fourth phase of my mathematical studies was not an attraction to math as much as a repulsion to engineering" [p. 3.4]. With the irony and humor that were typical of him, he writes,

> I also felt that since I had a block against mathematics I could either get psychoanalyzed or find out what mathematics really was. The second option being cheaper and more productive, I decided to get a doctorate in pure mathematics [p. 3-5].

It was then that his love for mathematics blossomed. He once again studied with Castelnuovo, this time in the theory of probability, and with Mauro Picone in numerical analysis. He studied nuclear physics and thermodynamics with Enrico Fermi [Salvadori 1997, 23; 1987, 57]. He mentions that he studied with Federigo Enriques (1871-1946), but doesn't give specifics. His first idea for a doctoral thesis was an application of mathematics to an engineering problem involving dams, but Levi-Civita discouraged him from this [p. 3.10]. His doctoral thesis, under the direction of Mauro Picone, concerned the calculus of variations [p. 3.12].

Salvadori's description of his gargantuan efforts to complete the thesis is as charming as it is telling. With a draft in hand, he went to Selva di Cadore in the Dolomites during August holidays in 1933 to consult with Picone in order to finish it:

> In the ten days I spent there with Picone I got the best lesson in mathematical rigor and the most painful humiliation of my entire life. It seems that the proofs of <u>all</u> my theorems, of which there were twelve, were wrong. With almost sadistic pleasure Picone found holes in each one of them. Not irremediable mistakes negating the value of the thesis as a whole, but minor inaccuracies, small non-sequiturs, little stumblings in logic, which he fixed easily enough, but not without first rubbing-in his amazement at my amateurishness. I panicked because of my stake in finishing the thesis on time, but most because I needed his esteem for the future of my career. [...]

After ten days of mathematical agony I was on the brink of a nervous breakdown, when a miracle occurred: the proof of my last theorem was wrong, of course, but Picone could not find the right proof. I was elated and terrified [p. 3.13].

Salvadori worked through September and into October to find the one example for which the theorem was true. The insight necessary to finish the thesis came in a flash around the middle of October, and thus he earned his Ph.D.

Salvadori was under particular pressure to finish the thesis, because he had to be in London on 1 November 1933:

I had just won an international competition a fellowship that was going to pay for my stay at University College, London, to do graduate work for an academic year under professor E.G. Coker, the inventor of a practical method for evaluating stresses in structures by means of polarized light, photo-elasticity. The fellowship had been granted originally by the League of Nations, but at the time the League had taken the unprecedented action of "sanctioning", that is, of siding against Italy because of her undeclared and bloody war in Ethiopia, and Mussolini had forbidden the acceptance of such grants by Italian citizens, offering instead identical grants from the Fascist government [p. 3.14].[2]

This is perhaps the first hint of how political events would crucially shape Salvadori's decision to leave Italy for the United States. It was during the eight months in London that Salvadori became aware of how the outside world viewed Italy's Fascist dictatorship, and just how dangerous Mussolini himself and his regime was for Italians. He also met and talked to German refugees fleeing from Nazism, and their stories made a deep impression. But he was generally lonely in London and looked forward to returning to Italy. His future there looked particularly rosy: he could look forward to two part-time positions that would further his career, one with the Istituto per le Applicazioni del Calcolo under the direction of his former thesis advisor, Picone, and the other as assistant to his former professor in civil engineering, Giannelli.

Mauro Picone's IAC

Picone is universally recognized as the founder of a flourishing Italian school of mathematical analysis and as an exceptional organizer. Among other things, it is to Picone that we owe the autonomous development of numerical analysis and of automated calculation (one of the first computers in Italy was that installed in Rome in 1954 and inaugurated the following year).

Sicilian by birth, Picone studied first in Parma and then in Pisa, where he earned his degree in 1907 at the Scuola Normale Superior, which had been revitalized after Italy's political reunification in the second half of the nineteenth century. A student of Ulisse Dini (1845-1918), a mathematician famous for his rigorous purism, Picone received his illumination "on the road to Damascus" during the course of World War I from 1915-1918, an experience which radically changed his way of "viewing mathematics". In particular, the elaboration of new gunnery tables for the artillery – which he achieved by adapting the old tables developed by Francesco Siacci (1839-1907) to the particular geographic conditions of the region around Trento – lead him to write in his autobiography, "... you can imagine, after this success achieved with mathematics, the kind of new light in which it appeared to me. I thought: so Mathematics is not only beautiful, it can be useful as well."

After the war, Picone taught at the universities of Turin, Cagliari, Catania, Pisa, Naples and, finally, Rome, becoming known even at an international level. It was in Naples that, in 1927, he founded a small "Istituto di Calcolo", which would then follow its founder to the capital in 1932, to become the Istituto Nazionale per le Applicazioni del Calcolo (*INAC*). The adjective "National" was dropped in 1969, when the *IAC* was officially named after its founder.

Salvadori described Picone's founding idea thus:

> Picone, a vivacious, bright and aggressive Sicilian had been the first mathematician in Italy to recognize the importance of numerical analysis, a branch of mathematics concerned with approximate calculations and requiring only the use of the four elementary operations of arithmetic. A simple example will explain the gist of numerical analysis to the most unmathematical person. Let us assume that we wish to compute the square root of 5 (a high-school nightmare) and do not remember how to take square roots. Numerical analysis starts by noticing that 2 is not a bad approximation for the square root of 5 since the square of 2 is 4, a number fairly near 5, but smaller than 5. It is then noticed that since 5 divided by 2, that is, 2.5, is too large an approximation (the square of 2.5 is 6.25), the average of 2 and 5 divided by 2 will probably be a better approximation. The average of 2 and 2.5 is 2.25 and its square is 5.0625. We are almost there, but 2.25 is still a bit too large. If one wants <u>ever</u> better approximations of the square root of 5, one must simply continue taking the average of the last approximation and of 5 divided by it. In our example, at this stage, we take the average of 2.25 and of 5 divided by 2.25 and obtain 2.236 whose square 4.9997 differs from 5 by only 0.0003, that is by 3 hundredths of 1% and has been obtained by adding and dividing numbers only.

> When Picone became interested in numerical analysis the electronic computer was twenty-five years into the future and the pure mathematicians at the University of Rome despised the subject, although the Germans had already given it high status in some of their universities. With admirable single mindedness and great political skill Picone succeeded in getting appointed to a chair of mathematical analysis in Rome, left Naples University together with his brilliant assistant Miranda and, besides teaching the kind of math acceptable to his illustrious colleagues, obtained a small grant from the newly established Italian National Research Council to start in a small apartment in Rome in a new section of town the high-sounding "Institute for the Applications of the Calculus." (The words "numerical analysis did not appear in the name of the institute and for good reasons [p. 3.10-11].

The *IAC* immediately became an significant and original force in the Italian scientific panorama. It was the first time that mathematical research had been organized outside the rigid academic circuit; it was the first time that young researchers were directed through a channel that led to a considerable number of jobs (in relation to pre-existing conditions); it was the first time that Mathematics became both the subject and the object of consultancies, opening the way to new professional relationships. The changes that Picone was able to effect were not limited to structural and organizational aspects but involved the contents of research and the very meaning of the terms used when we say that we want to undertake and resolve a mathematical problem. A new numerical mentality had appeared on the Italian mathematical horizon. It was no longer sufficient to prove an existence theorem, and ultimately its uniqueness, but rather it was necessary – in an equally essential way – to articulate constructive procedures for calculating the solution. In other words, it was necessary to give the same attention and the same rigor to the determination of a numeric algorithm, for the proof of its convergence and for the margin of error of the approximation. All of this was accompanied almost naturally by a great attention to automated calculation, at first analogical and then, after the second World War, electronic.

Today the *IAC* is one of the CNR's two main centers of research in applied mathematics. The second center is the *IMATI*, the Institute for Applied Mathematics and Information Technology, with offices in Genoa, Milan, and Pavia. Since 2002 the *IAC* has had offices in four cities – Rome, its historic headquarters, Bari, Florence, and Naples – and now has a staff of some fifty-five researchers, plus students, fellows and external collaborators. Since its inception its object has been to develop mathematical, statistical and computer methods for resolving problems of great social and industrial relevance, within an interdisciplinary context.

Picone and Salvadori

It was precisely regarding the field of automated calculation that Picone turned to Salvadori. Taking advantage of Salvadori's presence in London, who had gone to there to for post-doctoral study in photoelasticity, Picone entrusted him with two projects, as described in a letter of 3 May 1934:

> Dear Salvadori,
>
> With the commissions that I am entrusting to you with this letter begins your involvement with this Institute, as I have long desired.
>
> First commission
>
> The Cambridge Instrument Company (45 – Grosvenor Place – London SW1) is the builder of the Mallock machine, which resolves systems of linear equations. The acquisition of such a machine by the Institute, given its elevated cost (some 100,000 Italian lire), must be preceded by a conscientious examination of the machine itself and by testing. I am entrusting all of this to you. I beg you therefore to go to this company, which has been informed of your visit, and to examine the entire functioning of the machine and to understand as much as is possible of the practicality of the operations necessary to guarantee its working as it should. As regards the testing of the machine, the best way is to make it resolve systems of equations to which we already know the solutions; attached to

this letter are two such systems, which also have the characteristic of symmetry, which is the discriminate of a quadratic form defined as positive.

Once the two systems have been resolved you can send us the solutions which we will compare with our own.

Second commission

We have been asked to trace isostatic lines relative to the equilibrium of a rhombus-shaped plate with the acute angle measuring 60° supported on two opposite sides and subject to a uniformly-distributed load. Now we are wondering whether if our formulae for approximations are sufficient and I pose to you the question: is it possible to measure isostatic lines photographically using photo-elastic methods? If it is possible I urgently beg you to involve the institute where you are studying in this problem, which is of extreme importance to us.

Of equal interest is Salvadori's answer, given by return post in a letter dated 7 May 1934:

Most dear Professor,

At the same time that I thank you for the trust you have shown me, I hasten to tell you the results of my visit to the Cambridge Instrument Company.

I received your letter this morning and discussed the matter with Prof. E.G. Coker, director of the laboratory where I work, who introduced me to the director of the department of statistics of our College, Prof. Pearson.[3] Prof. Pearson does not know the Mallock machine, but has written me a letter of introduction to Mr. Whipple, one of the general directors of Cambridge Instrument. I went to that company, and in the absence of Mr. Whipple, spoke to his assistant.

The Mallock machine is the fruit of a collaboration with Prof. Mallock of Cambridge and Mr. Mason, general director of Cambridge Instrument; there exists a single exemplar, in the possession of Mr. Mallock, in Cambridge; it is not a machine that is easy to market, indeed one for the Institute for Calculation would be the second one ever built. I cannot therefore have, as I had hoped, information from a body that has been in possession of the machine for some time. I had the systems of linear equations that you sent me sent today to Cambridge; Mr. Mason will resolve them as soon as possible, and I will probably go personally next week to Cambridge, so I can have an accurate explanation of how the machine functions by Mr. Mason or Mr. Mallock. Further, I intend to personally resolve at least one of the two systems so that I have an idea of the rapidity and the greater or lesser simplicity of the operations. As soon as the values obtained are available I will send them to you.

The price of the machine is 1,725 pounds (equal to 103,500 Italian lire), with consignment in any English port; the expenses of packing and customs fall to the buyer. This price, according to Cambridge Instrument, barely covers the cost of manufacture.

I await any further instructions before going to Cambridge.

Thus began, with a commission of great trust, the consulting activities of Mario Salvadori for Mauro Picone and for the *IAC*.

Once back in Rome, at the *IAC* under Picone's direction, Salvadori worked on mathematical problems with practical applications for industry, state and science, applications that were looked down upon by the pure mathematicians at the universities.

> The only calculators available to us then were mechanical and hand cranked. Later we were able to buy a few electrically powered mechanical calculators, which were a little faster and less noisy. Yet by 1935 Picone's vision and petsistence had given the institute, by now the <u>National</u> Institute for the Applications of Calculus, the entire top floor of the new palace of the National Research Council and enough money to pay (miserly) a staff of thirty. As engineering consultant, a part-time job that took half of my time, I made 600 lire or $30 a month. On the morning of the inauguration of the new headquarters we were requested to be at our desks at 8 a.m. When "il Duce" strutted along the corridor of the institute we cranked our machines by hand as fast as we could after setting all the levers of our calculators at the 9 positions, because in this configuration the calculators made the biggest racket. Il Duce could not miss the enormous significance to the future of the Fascist Empire of so many 999,999,999s being so loudly multiplied by 999,999,999. In a sense the bombast of these multiplifications was an honest representation of the empty racket of most Fascist activities. As I vigorously turned my crank I saw from the corner of my eye paunchy Mussolini in his black uniform followed by paunchy Picone in his black shirt. As soon as "he" left the racket stopped and we went back to our serious pioneering work [p. 3.11].

Salvadori's other part-time position was assistant to his former professor, Giannelli. He was less fortunate in his choice of mentor in Giannelli than with Picone. Giannelli was fearful of seeing his assistants outstrip their master, and he purposely foiled Salvadori's progress by slowing down publications by him and by delaying his attainment of the qualification of *libero docente* (a qualification that permits one to teach at university level without a permanent position as professor), which he in any case obtained in 1936 [Salvadori 1987, 59]. But finally Giannelli asked Salvadori to take his place for a lecture, and Salvadori discovered that he not only delighted in lecturing, but that he had a real talent for it. The subject of that first lecture was on the Hardy Cross distribution method [cf. Eaton 2006] for designing buildings in reinfotced concrete [p. 16.23]. In just a short time, Salvadori would meet Cross himself during a trip to the United States during which he presented the work of the *IAC* to colleagues and institutions on the other side of the Atlantic.

Rising tension in pre-war Italy

In 1938, Giannelli had a new reason to keep Salvadori at a distance. On 17 January, a statement by the Fascist government appeared in all papers claiming that there was no Jewish problem in Italy but reserving the right to limit activities of the Jews as they saw fit. One was considered Jewish if both parents were Jews, or if there had been overt Jewish behavior (such as marrying according to a Jewish rite, or belonging to a Jewish community). Salvadori's mother was Jewish, but his father was Catholic, so according to "the laws" he wasn't a Jew. But he was in fact a member of the synagogue, and had been married in a Jewish ceremony. He might not have to have been afraid, but he was. Recalling what he had heard in London from the German Jews fleeing Hitler, he says, "if I had been horrified then, I was terrified now" [p. 21.3].

At first the situation was so unclear that no one knew quite what to do. The question of precisely what kind of activities might be limited was secretly discussed among Jews. The first racial laws appeared in September 1938; these were progressively reinforced as the year went on. Salvadori saw his dream of being a university professor slipping away. To some of those close to him, this seemed almost a minor problem:

> My in-laws were so blind that to ignore the obvious they did not even have
> to bury their heads under the sand. There was nothing to fear; at most I
> might not be allowed to teach at the University. And so what? I was an
> engineer, wasn't I? I could be a professional man ... [p. 21-4].

The laws continued to tighten restrictions on the activities of Jews: on 2 September all Jewish high school and university professors were dismissed from their posts. Within a few days, Salvadori was involved twice in racial queries. First he received a telephone call from an embarassed Picone telling him that he should stay away from the *IAC* for a few days as his "Aryanity" was being investigated [p. 21.10]. Next he received notice regarding his position at the university as Giannelli's assistant, and again his position was suspended until his racial status could be determined [p. 16.23-24]. Not without misgivings and apprehension, and to the consternation of his family, Salvadori resolved to leave Italy as soon as an opportunity presented itself.

That opportunity arrived when Salvadori took part in a competition for a grant to study the organization of television in either the United States or Great Britain so that the Fascist government could organize their own system [p. 20.57]. One of the three members of the competition commission was Enrico Fermi; Salvadori credits Fermi with intervening on his behalf [Salvadori 1987, 59]. He was able to obtain six-month tourist visas for himself and his wife, Giuseppina, effective 16 September, and booked passage on a steamship that sailed six days later. He also proposed to Picone that he officially represent the *IAC* at the fifth Congress of Applied Mechanics to take place held in that same month in Cambridge, Massachusetts, an offer that Picone accepted. In the historic archives of the Institute there exists a report that Salvadori sent to Picone on 15 November 1938. This is a very important document, because its official tone and account of events is very different from the account of the same events given in his autobiography:

> Having decided to go to the United States with a grant from the Institute
> for Cultural Relations with Foreign States with the object of studying the
> American organization of television, Prof. M. Picone, Director of the
> National Institute for Applications of Calculation accepted my proposal of
> undertaking promotion of the Institute in the spheres of the universities
> and industries of the United States. To this end I proposed a presentation
> at the Fifth International Congress of Applied Mechanics in Cambridge,
> Mass. (USA), on the activities of the National Institute for Applications of
> Calculation, a report that was written by me in English and which was one

of the papers accepted for publication in the proceedings of the Congress by the directive committee, which in April declared its enthusiasm for the initiative taken and held that the report would have been received with the greatest interest.

Some difficulties that arose when it was time to obtain the authorization for exporting currency from the Institute for Exchange, prevented me from leaving Italy in time to participate in the Cambridge congress and read the report relative to the Institute at the congress itself. In spite of this, the paper was transmitted to the directive committee, which will see to its publication within the first months of 1939.

As things stood, I was able to leave Italy on 22 September on the steamer Rex and reached New York on 29 September. On 30 September I submitted the participation fee for the congress to the directive committee, which gave me the right to publish the report, letting it be known that I would go to Cambridge personally to tend to the details of the publication.

Because I was without a letter of introduction, I thought it opportune to go at once to the American Society of Civil Engineers in my position of engineer and university professor, with the aim of making contact with the principle New York universities via the most important engineering society. I had the pleasure of meeting, in the absence of the Secretary, the Vice-Secretary General Eng. C.E. Beam and the Vice-Secretary Prof. Eng. A. Richmond, professor at Columbia University, to whom I expressed my wishes after having illustrated the aims of the Institute. The very cordial reception on the part of the Secretaries of the A.S.C.E, who were enthusiastic about our organization, permitted me to perform all of my successive activities profitably.

Eng. Richmond immediately arranged an appointment for 3 October for me with Prof. Eng. R.D. Mindlin, professor of elasticity and photoelasticity at Columbia University. This university, the most important in New York and one of the largest and most important in the United States, educates 40,000 students and has a teaching staff of 3900 professors and assistants.

Prof. Mindlin was pleased to invite me to lunch in the name of the university and was extremely interested in my exposition of the activities of the Institute, declaring that to his knowledge there were no such institutes in the United States and asking many for many details about calculation techniques and problems treated by the INAC. I also learned that the name of the INAC and my own were very well known as a result of the pre-publication of the proceedings of the Congress mentioned above and that the fact that I was unable to make my presentation was much regretted. I was then invited by Prof. Mindlin to visit his laboratory of photoelasticity as well as the laboratory for material testing, whose Director was also very interesting in the object of my visit. Finally, I had the honor of being introduced to the Dean of the Engineering School, Prof. J.K. Finch, and to Prof. J.M. Garrelts, professor of structural mechanics, to whom I briefly explained the activities of the INAC.

The next day Prof. Mindlin wanted to me to come again to Columbia University, as a guest of the university, in order to have lunch with Prof. L.P. Siceloff, professor of calculus there, who was very enthusiastic about the INAC and asked for specific information regarding the mathematical methods we use. I gave both Prof. Mindlin and Prof. Siceloff a complete series of the notes of the INAC that I had brought to America, receiving in exchange their own interesting publications.

The report then continues with descriptions of similar meetings with department heads and professors, many of whom were internationally known, at other universities in the United States, including Princeton, Yale, and Harvard. One such meeting must have excited him greatly, for at Yale he met the engineer Hardy Cross, whose distribution method for designing structures in reinforced concrete had been the subject of his first-ever university lecture, as mentioned earlier. The report describes the meeting thus:

Yale University enlivens the city of New Haven with its characteristic colleges in the English style, in which students live, going to study in the various faculties. Prof. Hardy Cross, one of the most noted American cultivators of Science of Construction and author of a famous method for the resolution of hyperstatic frame systems which I myself was the first to make known in Italy in 1933, has taught Science of Construction at Yale for a little more than a year, being called there from the University of Illinois. He received me in his studio on 10 October for a lunch in the company of Prof. Krinine, professor of the theory of foundation soils, and his assistant, Dr. Palladino, of Italian origin, and showed me the most cordial hospitality. First of all they presented me to the Dean of the Faculty of Civil Engineering, Prof. S.D. Dudley, to whom I explained the purpose of my visit and who told me that in the name of the University he was honoured to meet a representative of Italian Universities and the Institute of Calculation, for which he had great admiration both as organizations and for their results, and saying how very sorry he was that a previous engagement prevented him from lunching with us.

I visited the Faculty buildings, and was treated to lunch at the University Club, during which I briefly described our methods of calculating. Prof. Cross very much appreciated the contributions made by the l'Istituto Nazionale per le Applicazioni del Calcolo (INAC) to pure and applied science and demonstrated to me two problems, the first connected to his method of calculating frames, and the second to the calculation of arches in reinforced concrete, to which he had given his attention without, however, achieving conclusive results, given that they were of a specifically mathematical nature. I proposed to Prof. Cross that we place the problems in the hands of the INAC and he was very enthusiastic about this collaboration with Italian scientists.

Dr. Palladino was courteous enough to escort me to visit the buildings of the main colleges, and then to take me back to Prof. Cross, who wished to introduce me to his class of graduate students, presenting me with many flattering words about the Institute. Naturally I left with Prof. Cross a complete series of notes from the Institute, receiving in return copies of his interesting publications.

The report then includes a description of a lecture that he was invited to give at 8:30 on the evening of 27 October at Columbia University, preceded by a dinner in his honor, just two days before his departure for Italy:

At dinner Prof. Mindlin, Prof. Garrelts, Prof. Biot and Prof. Nervmann were present. Prof. Siceloff introduced me in the name of Prof. Finch, who was away from New York, to an audience of more than 50 people, some of whom were professors, others of whom were professionals. My lecture dealt with the same material that had formed the object of my paper for the Cambridge Congress, complete with many details and extended in scope, since the communication for the congress had had a limit on length.

At the end of the lecture, which lasted more than three-quarters of an hour, I had the opportunity to respond to numerous questions of technical and general nature by about ten members of the audience, and to distribute the remaining copies of the publications of the Institute that I had reserved for the occasion. A professor of chemistry at Columbia University wished to pose to me an important query regarding mathematics applied to chemistry, and begged me to subject it to the study of the INAC. Among others, present at the lecture were two of my former students at the University of Rome (Drs. Ripetto and Calandra), as well as two professors of Italian origin at City College of NY. The lecture was concluded with thanks by Prof. Garrelts in the name of the School.

On 28 October, the day before my departure, I was introduced to Prof. J.W. Barker, Dean of all the schools of engineering at Columbia University, and to Prof. Pegram, Dean of the School of Physics and of that specialization of the same university. These two personages as well demonstrated keen interest in the work of the Institute, during a cordial interview.

Salvadori concludes his report with six points:

a) the publication of the paper presented at the Cambridge Congress will make the work of the INAC known to the American scientific-technological world, which demonstrated during the Congress the keenest interest in our organization;

b) the promotion made personally by the undersigned in academic circles aroused the most unconditional admiration for the informing ideas and the practical actuation of our Institute, as is clear from the statements of no less than 27 professors, some of which are very well-known academic personages of international acclaim;

c) there does not exist in the United States any specific organization with the aims of the INAC, as I was told repeatedly in all the scientific and academic milieus that I frequented, even though individual mathematical consultants for industry and some institutions can in very particular cases fill this need;

d) even though the American industrial organization is partially equipped for the theoretical resolution of many problems and admirable equipped for their experimental resolution, a collaboration between the INAC and American institutes and industries would be possible. One fruit of this brief visit was the beginning of this collaboration in the scientific arena;

e) the exchange of publications between the INAC and the American institutes should favor the acquaintance with the Institute on the part of American technological-scientific organizations;

f) the expressions of sincere and lively admiration of the INAC on the part of many and diverse persons demonstrates in an evident way that the Institute by now enjoys a well-deserved fame of an international nature, which reflects the glory of the National Council for Research and Fascist Italy.

The closing remark of this letter shouldn't be surprising: this is an official, written report at a moment in which the personal situation of many Italian Jews was unclear. But the description in the official report is a far cry from Salvadori's account of it in his autobiography.

In the official report, he describes meeting Raymond Mindlin, but doesn't mention that Mindlin was also a Jew to whom Salvadori confided his predicament, and that Mindlin was ready to help Salvadori leave Italy permanently and set up a new life in the United States. Mindlin and Salvadori established a code by which Salvadori could communicate with Mindlin by letter after returning to Italy. If the situation was as dire as Salvadori feared it might be, he would write to Mindlin saying that his illness was worse and couldn't be cured in Italy, and if Mindlin was able to help, he would answer Salvadori saying that medicine was available in the United States and that Salvadori should come over for a cure. After having taken the precaution of hiring a safe-deposit box in a bank, leaving all of his wife's jewels in it, and giving a copy of the key to Mindlin, Salvadori returned to Italy to see which way the wind was blowing. Even the purpose of the lecture he was invited to give at Columbia two days before his departure was not so much intended to inform the audience about the activities of the *IAC* as much as it was for Columbia professors to judge Salvadori's qualifications:

I was unaware of the fact that, in a country in which professors are not chosen by competition, seminars are often used to test the qualifications of a person aspiring to join the faculty. Ray had organized the meeting very skillfully and I don't believe I let him down [p. 21.24].

Indeed, the situation was as bad as Salvadori had feared. An official notice of temporary suspension from the *IAC* pending verification of his racial status arrived in November 1938 [p. 21.27], just a few days after his return from the United States, during which visit he had acted as an official representative of Italy and the *IAC*. Salvadori wasted no time. He wrote the official report for Picone about the contacts that he had made while acting as representative of the *IAC* to industries and universities. He wrote the official report for the Ministry of Communications about his meeting with the president of the Radio

Corporation of America, and he ascertained that the visas granted on 16 September were still valid for re-entry to the United States. After that, he gathered what he could to take with him and said good-bye to friends and family.

Salvadori and his wife arrived in New York on 13 January 1939 to begin a new life. The very next day, a wire reached him saying that his suspension due to racial reasons had been rescinded. He wrote, "It was my incredible luck to receive such news the day after my arrival in New York" [p. 21.32].

Overseas contact

Almost a year would pass before Salvadori contacted Picone. In a letter dated 5 November 1939 he wrote:

> I am truly sorry and wish to apologize for not having resume contact with you until today, but first of all the confusion of life in America, and then my illness have physically kept me from doing so. My father has kept me regularly informed about the progress of my business with the Council (CNR), and has recently sent me the letter of dismissal from the Secretary General.
>
> By uncanny coincidence this letter reached me on the same day in which I began work as an engineer in an important factory in New Jersey, but even this does not prevent me from confessing my disappointment at being removed from your Institute for which I have worked for four years, and from which I have learned so much and to which I hope that, within the modest limits of my capabilities, I have given what has been asked of me.
>
> But above all I regret that I must leave you, to whom I owe so much, and my dear colleagues.
>
> As you yourself have observed my situation has not followed the course that was shown and promised to me. The head of personnel of the CNR has not held my position sufficiently transparent, and it would seem that he has seen fit to dismiss me from the Institute because of my trip, in spite of my explicit declaration that I would return within the month of March and your disposition to await me until April.
>
> All told perhaps it is better that this has happened, because my reception in New York and the offer of the position that I currently hold have contemporarily made me decide to stay in the United States. The interesting nature of the work in a large American factory, with a starting salary that is four times what I was earning in Rome and is rapidly increasing, the vast possibilities of this country, combined with the not total acceptance of who I am, due to occurrences beyond your control and mine, would have forced me to tender my resignation, though I would have done so with regret. I am therefore grateful that the Secretary General found the best solution of all.
>
> In spite of my occupation of an industrial nature, I am still in contact with the various universities and would very much like to receive regularly the publications of the Institute, so that I can continue the promotional work

that began so well. If I could receive news about you personally I would be most happy.

I beg you to give my most cordial greetings to all the colleagues of the Institute, and to tell them of my good news, and for yourself please accept this expression of fond gratitude from your devoted student.

The occupation "of an industrial nature" to which Salvadori refers was with the Lionel Train Company, the largest American manufacturer of toy trains. By the 1940-1941 academic year he had already been assigned a course in "applied mathematics for engineers" at Columbia University.

Picone solicited him to establish scientific relations between the Department of Civil Engineering" of Columbia University and the *IAC*. But the second World War interrupted every possibility of this happening. It is significant, however, that at the end of the war, in 1945, Salvadori remembered his former teacher and the *IAC* and sent to Picone in Rome, via his father Riccardo, the following communication in a letter dated 7 November 1945:

> We also want to develop an electric machine to solve problems of vibrations and another, of which a small version has already been constructed by a mathematics professor at Columbia, to solve linear algebraic equations. This man got the idea of solving the equations with a machine that is so simple and so inexpensive that any ham radio operator could build it at home ... and he did so by means of an idea that is simply delightful. If it functions for many equations as simply as it does for now for four, the problem that science has been posing for many years ... will be resolved ... and with very little money, so that I would be happy to write a report in Italian so that you could build one over there.

But at that time Picone was already involved in one of his most important projects, that which, following the liberation of Rome in 1944, would engage him in the attempt to bring Italy and the *IAC* to the level achieved by the great industrial powers in the field of instruments for calculation. Picone's long campaign for the construction or acquisition of the "first great Italian electronic calculators", which concluded in 1954, represents one of the few successful attempts in Italy in the direction of scientific and technological innovation.

Picone had learned of the construction of the first American computer from an article dated 4 August 1944 in *The Stars and Stripes*, the newspaper of the American armed forces in Europe. He immediately began to campaign for one, but the post-war political conditions – the beginning of the period that we usually call "the cold war" – were such that only in 1954 could his dreams become reality, when with a contribution from the funds provided by the Marshall Plan he was able to acquire a computer from the firm named Ferranti in England. During this decade, Salvadori assisted Picone with assiduous, caring, and intelligent advice, in exchange for which the *INAC* carried out the execution of various calculations that were helpful in promoting Salvadori's image as an engineer-mathematician. It can safely be said that Salvadori's role of ambassador was a significant factor in realizing the dream to which his former professor rightly attributed such importance. Picone himself attests to this in his address during the 1955 inauguration of the commencement of service of the FINAC, as was called the Mark I computer that

Ferranti constructed purposefully for the Institute (the acronym is a combination of F for Ferranti with the *INAC* of the Institute):

> Once peace was established after the last war, news arrived in Italy, fragmentary and imprecise as the reports were, of great calculating machines, of immeasurable power, which had been constructed in the United States during the war for military purposes. Thus was born in my soul, and in those of my colleagues, the hope that we would come to own one of those machines, so that we could obtain concrete solutions to all the problems that awaited them.

> From that moment we have tenaciously pursued that goal, and after many missions abroad – facilitated by cordial receptions wherever we went – after having tested the majority of the new calculating machines, whose construction in the meantime was being undertaken and perfected in several of the most technologically advanced nations, we were finally able to select and obtain for our use the powerful electronic calculator that we today solemnly inaugurate, in the esteemed and heartening presence of the Head of State (...)

> Our experiences thus far with this calculator have deeply convinced us that the considerable expense sustained by the State for its acquisition will most certainly, in the right hands, be amply compensated for by the benefits that its well-planned employment will produce in terms of the scientific and industrial progress of our Nation.

With the arrival of the computer, the already flourishing collaboration between the *INAC* and Mario Salvadori was strengthened, and Salvadori could propose to the *INAC* forms of collaboration that extended to other colleagues and institutions in the United States. Thus for example in a letter of 9 October 1956, Salvadori wrote:

> Dearest Picone,

> I am so sorry to have only seen you for such a short time last summer, but I am counting on being there again next summer and to talking to you at leisure.

> The American Society of Civil Engineers has informed me that they could send the Institute 450 copies of my published report on the critical loads of fixed beams for $140.00. I would be grateful to you if your could let me know if you want them, or given the cost, if you would want a smaller number.

> I came back to New York two weeks ago and have resumed my work. An engineer colleague of mine, who works on thin shells, has asked me to ask you how much it would cost to solve a system of 100 equations with 100 unknowns (linear, naturally). I don't believe that this constitutes an academic research, and therefore the Institute must be compensated. I would be grateful if you could send me an answer.

A final decision

In spite of his success in the United States both as a professor at Columbia University and as a practicing structural engineer, Salvadori and his wife struggled to decide whether or not to return to Italy now that Fascism no longer threatened and a lasting peace had been restored. Picone himself had solicited Salvadori about this possibility several times. In a personal and confidential letter dated 6 April 1957, Salvadori expressed both his desire and his difficulties to his former teacher. That by this time their relationship had outgrown the limits of student and teacher and had become one of affectionate equals is evident from the use in the original Italian of the informal "tu" instead of the formal "Lei" that Salvadori had formerly used to address Picone. He wrote:

> We are counting on leaving for Italy on 12 June and hope to see you as soon as we reach Rome, towards the end of June, but before then I hope to be able to ask you some questions of a general nature, of the kind that one can only ask of a beloved teacher whose friendship is precious.
>
> As you know, I have been a full professor for some years at Columbia University and further, for the past two years I have been assigned to teach courses in structural mechanics at the school of architecture at Princeton University. My professional practice has been going strong for the past two years, since I went into partnership with a brilliant American engineer,[4] with whom we design daring structures for the best architects here, and undertake research in the dynamics of structures for the American Government. Thus I couldn't be more fortunate in my American career, from an academic and professional as well as from an economic point of view. But you know me and you know my family, and it won't surprise you if I tell you, naturally in the strictest confidence, that in our heart of hearts Giuseppina and I still hope to return to Italy. On the one hand I would be loathe to return if I had failed in the United States, and on the other hand the desire to return has been tempered by the success I've had here. This is why, in spite of the affectionate encouragement recently, I had never determined to open up to you, which I now do in hopes that you will have some fatherly advice for me.
>
> I just turned 50 a few days ago and know, or believe I know, that teaching is probably out of the question for me in Italy, even though I am still enrolled in the register of teachers of the University of Rome. You know that for many years I have continued researching in mechanics and structures and that I have some 60 publications to my credit, plus 3 books on applied mathematics, of which the first was published in Italy as well and the second (on numerical calculus) translated into Russian, Portuguese, and Chinese (although this last is not yet published). I believe that you know that I speak 5 [languages] and Romanian. And now here is the question: what kind of work and occupation could I find in Italy, if I decided to return?
>
> It is obvious that I cannot accept some minor position, and that I wish to perform an activity that gives me deep satisfaction, in addition to a dignified income. I am too aware of my very modest worth to add that I

wish to make a contribution to Italian scientific life, but I do not hide that a position in which my knowledge of the international scientific world could result in even a small contribution to our nation would attract me more than any other.

I am completely in the dark regarding the regulations of universities and public and semi-public institutes, and I have no idea what industry might offer me. I think because of your knowledge and contacts with the scientific and industrial spheres and because of your affection for me, and because of your wisdom, you are the only person in Italy to whom I can turn and who can give me dispassionate and caring advice. I thus await your answer, and wish to express now my profound gratitude.

In his answer Picone suggests several solutions that range from possible rehiring at the *INAC* in his former role as "consultant" to participation in a competition for a chair in structural mechanics at the University of Palermo. Salvadori opted for this last, and participated in the competition, qualifying for the position (along with two other Italian colleagues). He hoped to be called to the University of Florence, where in 1959 a position was supposed to become available. But this hope was not to be realized because Salvadori had lost his Italian citizenship in July 1940 (as we are informed in a letter dated 17 February 1958). He therefore remained in the United States. But he was deeply disappointed, even if it didn't arrest his extraordinary capacity for work. Here is what he wrote to Picone in a letter dated 5 October 1958:

By this time you will have learned of the negative outcome of my competition: after having received the decree nominating me from the Ministry, it was taken away because I was unable to show that I was an Italian citizen on the day that I enrolled in the competition. Dr. Di Domizio, General Director of Secondary Education, was extremely nice during a long interview, but he could do absolutely nothing, given the Italian laws concerning nationality and the rules of the competition.

The dream of returning to Italy as a professor has thus vanished, after months of hope and anguishing doubts, and nothing remains for me now but to stay in America.

I still have the possibility of beginning a professional role as engineering consultant in Italy, and this I will begin to do. I have a consultancy with Olivetti concerning its mechanical calculators, and consultancies of a structural nature with groups of architects. Indeed, I have opened an office in Rome together with a group of American engineers, which I will tell you about when I see you.

I am now leaving for the United States, where I am going to give three lectures, in addition to a lecture in Madrid en route. I will be in Rome again on 21 October, and together with my family, will pass the whole year at Via Michele Mercati 10, Tel 870-295.

It was perhaps good for Mario Salvadori but a great loss for Italy that this is how things went. But this judgment is part of another story!

Acknowledgment

The authors wish to thank Michael Kazin, Mario Salvadori's stepson, for having provided us with a copy of the autobiography *A Tangential Life*.

Notes

1. Page references provided in the text for quotations from Mario Salvadori refer to this autobiography.
2. We believe that Salvadori makes a slight error here, anticipating the international isolation of Italy by some years.
3. Egon Sharpe Pearson (1895-1980), son of the well-known Karl Pearson (1857-1936).
4. Paul Weidlinger (1915-1999).

References

Eaton, Leonard. 2006. *Hardy Cross: American Engineer*. Champaign IL: University of Illinois Press.

Salvadori, Mario. n.d (1977?). *A Tangential Life*. Unpublished.

———. 1987. Ricordando Enrico Fermi (Memorie di un non-fisico). *Il Nuovo Saggiatore* 1, 3: 54-65.

———. 1997. A Life in Education. *The Bridge* 27, 2 (Summer 1997): 21-27.

About the authors

Kim Williams received her degree in architecture from the University of Texas in Austin. She practiced architecture in New York City until 1987, when she moved to Italy. She is the director of the conference series "Nexus: Relationships Between Architecture and Mathematics" and editor-in-chief of the *Nexus Network Journal*. Her books include *Italian Pavements: Patterns in Space* (Houston, Anchorage Press, 1997) and *The Villas of Palladio* (New York, Princeton Architectural Press, 2003).

After beginning in physics, Pietro Nastasi earned his degree in mathematics in 1965 with a thesis on the application of Stieltjes integrals for the calculation of mathematical reserves in life insurance. His early scientific studies regarded positional geodesic astronomy and climatology in the area of Palermo. From the end of the 1970s, he has dedicated more and more of his research to the history of mathematics, a discipline that he has taught as Associate Professor since 1985 at the Faculty of Sciences of the University of Palermo. In this area of research, from his first studies on "local history" his principle interests have been progressively oriented towards more general themes and are presently focused on the study of institutional aspects of Italian mathematics after the unification of Italy. Among his studies are those dedicated to the repercussions of the racial laws on the Italian scientific world, and is the co-author (with Giorgio Israel) of *Scienza e razza nell'Italia fascista* (Bologna, 1998), and co-author (with Angelo Guerraggio) of *Italian Mathematics Between the Two World Wars* (Springer, 2006).

Federico Foce

Dipartimento di Scienze per
l'Architettura
Università degli Studi di Genova
Stradone S. Agostino 37, 16123
Genova, ITALY
fedefoce@arch.unige.it

Keywords: Milutin Milankovitch, masonry construction, arch theory, history of structural mechanics, masonry arch, line of thrust

Research

Milankovitch's Theorie der Druckkurven: *Good mechanics for masonry architecture*

Abstract. During the nineteenth century many studies on the theory of the thrust line were written in connection with the stability of masonry structures. However, a general treatment of the theory of the thrust line from both a mechanical and mathematical point of view may be found only in the contributions of the Serbian scholar Milutin Milankovitch, published between 1904 and 1910 and substantially unknown to the historians of mechanics applied to architecture. This paper aims at presenting Milankovitch's theory and discussing its improvements with respect to the previous literature on the subject.

Why good mechanics for masonry architecture?

There are different reasons for putting this question with respect to Milutin Milankovitch's contribution on the theory of the line of thrust applied to masonry vaulted structures. From our modern point of view, the result of Heyman's lesson, the first reason concerns his methodological approach. When Milankovitch discussed his doctoral thesis, *Beitrag zur Theorie der Druckkurven* [1904], (Fig. 1) at the Technische Hochschule in Vienna in 1904 and then published the two papers in the *Zeitschrift für Mathematik und Physik* [1907a, 1910],[1] the application of elastic theory to masonry and stone structures was a well-established trend. Starting from the 1870s the old tradition of studies based on the model of the arch as a system of rigid and infinitely resistant voussoirs was rapidly abandoned and the "new theory" of the elastic arch became the official tool of the structural engineer, as well as for systems and materials that only with difficulty obeyed the classical hypotheses of the mathematical theory of elasticity. This new attitude, a product of the times as well, was the inevitable result of nineteenth-century progress in structural mechanics related to the great development of the general methods for the analysis of hyperstatic elastic systems.

Milankovitch takes his distance from this official trend. He obviously knows that the only rational way for solving a statically indeterminate system requires adopting the elastic approach, that is, using the complete set of equilibrium, compatibility and stress-strain equations in order to determine the actual state of the system. His purpose, however, is not to search for the actual state of the arch. His interest is focussed on a general theory of the equilibrium of masonry structures and for this aim there is no need to take the elastic approach. It is sufficient to assume some basic hypotheses concerning the mechanical behaviour of masonry materials and to state the equilibrium equations in accordance with them. Now, these hypotheses are essentially three: 1) masonry has no tensile strength; 2) its compressive strength is practically infinite in comparison with the stresses occurring in real structures; 3) friction is large enough to prevent sliding between masonry elements.

Nexus Network Journal 9 (2007) 185-210
1590-5896/07/020185-26 DOI 10.1007/s00004-006-0039-9

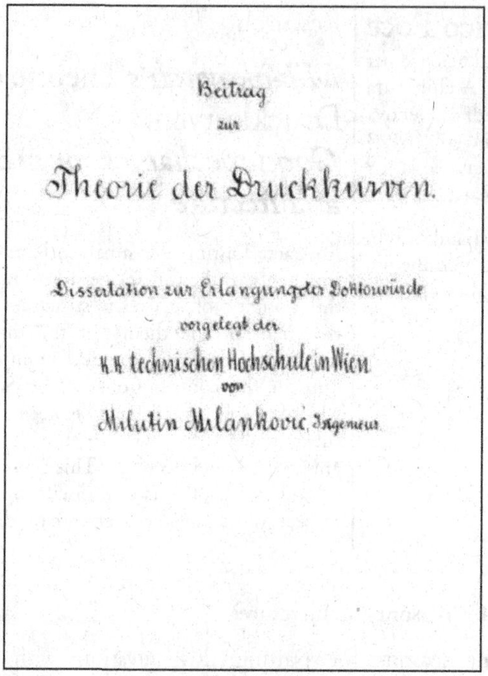
Beitrag

zur

Theorie der Druckkurven.

Dissertation zur Erlangung der Doktorwürde

vorgelegt der

k.k. technischen Hochschule in Wien

von

Milutin Milankovic, Ingenieur

Fig. 1. Frontispiece of the manuscript of Milankovitch's doctoral thesis [1904]

Milankovitch tacitly adopts these assumptions, nowadays called Heyman's hypotheses after the first essay that the great Cambridge scholar devoted to the stone skeleton [1966], and for this reason his work on the theory of the line of thrust is of particular value from a methodological standpoint. He recognizes that the main point of masonry structures concerns the global stability, not the local stresses. In this sense, and in spite of the times, his contribution ideally belongs to the pre-elastic tradition[2] and anticipates the present view on the matter.

A second reason comes from the scientific quality of Milankovitch's analysis. As far as we know, his theory of the line of thrust is probably the most general discussion in the technical literature on the subject. Among the previous and later studies on this topic it is be difficult to find a similarly high standard from both a mathematical and mechanical point of view, as we shall see in the following sections. The author himself seems to be conscious of that and his criticisms towards some previous contributions are particularly noteworthy in this sense.

Finally, a third reason concerns the role of the historical research in the field of the mechanical sciences. In spite of the remarkable level of Milankovitch's contribution, it must be said that for many decades it remained totally unknown to the historians of structural mechanics.[3] There are reasons also for that, and the main one is probably connected with the singular scientific career of the Serbian scholar. It is true that Milankovitch attended the Technische Hochschule in Vienna (today Vienna University of Technology) where he graduated in Civil Engineering in 1902 and earned his doctorate in 1904. After discussing his thesis and publishing the two papers cited above, he also worked in the then-famous firm of Adolf Baron Pittel Betonbau-Unternehmung in Vienna, wrote some technical texts on reinforced concrete [1905, 1907b, 1908] and built dams, bridges, viaducts, aqueducts and other structures in reinforced concrete throughout the Austria-Hungary of the time. However, this initial career as a structural engineer came to an end in 1909 when he was offered the chair of applied mathematics (rational mechanics, celestial mechanics, theoretical physics) in Belgrade and decided to concentrate on fundamental research in geophysics. As a matter of fact, Milankovitch's name is best known for his theory of ice ages, relating variations of the Earth's orbit and long-term climate change, now known as Milankovitch cycles.[4] It is clear that, in comparison with his main scientific production, the first studies on the line of thrust are forgotten episodes. The task of the historian, whatever his field of research, is to rediscover what time forgets and to select, among the many mediocre episodes, the few good ones. This is the case with Milankovitch's forgotten *Theorie der Druckkurven*.

The mathematische Stilisierung *of the mechanical problem of the arch*

We have already underlined the general character of Milankovitch's analysis. This character comes out of his correct *mathematische Stilisierung* – we could say mathematical modelling – of the mechanical problem concerning the equilibrium of an infinitesimal voussoir of the arch. The main steps of this analysis concern:

1) Location of the centre of mass of an infinitesimal voussoir

A first important point for a correct *Stilisierung* concerns the position of the centre of mass of an infinitesimal voussoir with respect to the middle point of the joints. In order to determine this position Milankovitch considers a finite voussoir $NN'N_nN_n'$, bounded by the joints NN' and N_nN_n', with centre of mass at point S (fig. 2).

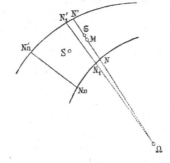

Fig. 2.

By moving the joint N_nN_n' to the position N_1N_1' infinitely close to the joint NN', the point S moves to a limit position G, intentionally drawn by Milankovitch on the line of the joint NN' and in this sense called *Scherwpunkt der Fuge NN'* (centre of mass of the joint).[5] The distance of this point G with respect to the middle point M of the joint NN' can be easily determined. Let $\overline{\Omega M} = \rho$ be the distance between M and the point of intersection Ω of the two joint lines NN' and N_1N_1', $\overline{NN'} = \delta$ the thickness of the joint NN', so that $\overline{NM} = \delta/2$. Moreover, let df_1 and df_2 be the areas of the infinitesimal triangles $\Omega NN'$ and $\Omega NN_1'$, whose centres of mass have the distances $\frac{2}{3}(\rho + \delta/2)$ and $\frac{2}{3}(\rho - \delta/2)$ from the point Ω, respectively. Then, the following equality holds:

$$(df_1 - df_2)\overline{\Omega G} = \tfrac{2}{3}(\rho + \delta/2)\,df_1 - \tfrac{2}{3}(\rho - \delta/2)\,df_2 .$$

Moreover, the ratio between the areas df_1 and df_2 is proportional to the ratio of the squares of the distances $\frac{2}{3}(\rho + \delta/2)$ and $\frac{2}{3}(\rho - \delta/2)$, so that

$$df_1 : df_2 = (\rho + \delta/2)^2 : (\rho - \delta/2)^2 .$$

From the two previous relations, it follows that

(1)
$$\overline{\Omega G} = \rho + \frac{\delta^2}{12\rho}$$

that is, the centre of mass G of the infinitesimal voussoir has the finite distance $\overline{MG} = \dfrac{\delta^2}{12\rho}$ from the middle point M of the joint NN'. This is the general case. The particular condition $\overline{MG} = 0$ is true only in the case of parallel joints ($\rho = \infty$) or in the case of vanishing thickness ($\delta = 0$).

2) General typology of geometry and load condition

The general feature of Milankovitch's approach concerns also the geometry of the arch and the load condition. As shown in fig. 3, the intrados and extrados lines are generic continuous (regular) functions, so that his analysis takes into account the case of variable thickness of the arch ring.

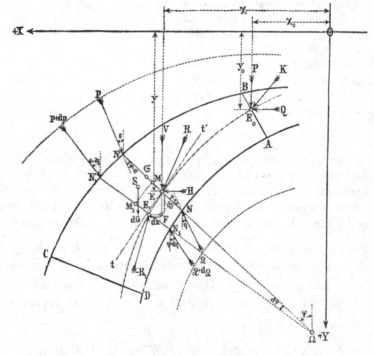

Fig. 3

Further, besides the self weight the load condition is represented by generic continuous (regular) functions p at the extrados and q at the intrados, with variable inclination measured by the angles ε and η with the y-axis. Finally, the direction of the joints is not necessarily perpendicular to the axis of the vault.

3) Correct formulation of the rotational equilibrium equation of the infinitesimal voussoir

On the basis of the statement concerning the position of the centre of mass G, Milankovitch derives the rotational equilibrium equation of the infinitesimal voussoir $NN'N_1N_1'$. By supposing that the line of thrust goes through the points E and E_1 at the joints NN' and N_1N_1', and taking into account the self weight dG of the voussoir and the forces pde, qdi acting on the extrados and intrados elements de and di respectively, the equilibrium equation about E_1 has the form:

$$(2) \qquad Vdx - Hdy + M_g + M_e + M_i = 0$$

where V and H are the finite vertical and horizontal components of the resultant force R at the joint NN' with infinitesimal lever arms dx and dy about E_1, so that the first two terms are infinitesimal quantities of first order; the other terms M_g, M_e, M_i represent the moments of the infinitesimal forces dG, pde, qdi about the same point E_1, respectively. As far as their values Milankovitch correctly observes that, because of the infinitesimal distance between E and E_1 and of the finite distance of the application points of dG, pde, qdi from both E and E_1, the three moments M_g, M_e, M_i are also infinitesimal quantities of first order and the lever arms with respect to E_1 may be taken equal to the lever arms with respect to E. Thus the moment M_g is given by:

$$(3) \qquad M_g = -dG\left(\overline{MG} + \overline{ME}\right)\sin\varphi = -g\beta\delta\left(\frac{\delta^2}{12\rho} + \xi\right)\sin\varphi\,\rho\,d\varphi \;,$$

where $\overline{ME} = \xi$ represents the eccentricity of the resultant R at the joint NN', g is the specific weight of the masonry, β is the depth of the vault and $\rho\,d\varphi = d\sigma$ is the length of the arc MM_1, so that $dG = g\beta\delta\,\rho\,d\varphi$. The other two moments M_e and M_i are, respectively

$$(4) \qquad M_e = -(pde)\overline{N'E}\sin(\varphi - \varepsilon) = -p\left(\frac{\delta}{2} + \xi\right)\sin(\varphi - \varepsilon)de$$

$$(5) \qquad M_i = -(qdi)\overline{NE}\sin(\varphi - \eta) = -q\left(\frac{\delta}{2} - \xi\right)\sin(\varphi - \eta)di \;,$$

where ε and η are the angles formed by the direction of the load p at N' and the load q at N with the vertical line, as shown in fig. 3.

By introducing these expressions in the equilibrium equation Milankovitch finds

$$
Vdx - Hdy - g\beta\delta\,\rho\left(\frac{\delta^2}{12\rho} + \xi\right)\sin\varphi\,d\varphi - p\left(\frac{\delta}{2} + \xi\right)\sin(\varphi - \varepsilon)de
$$
(6)
$$
- q\left(\frac{\delta}{2} - \xi\right)\sin(\varphi - \eta)di = 0
$$

Given certain boundary conditions, for instance that for $x = x_0$ it is $y = y_0$, $V = P$ and $H = Q$, then the values of the components V and H at the joint NN' become

(7)
$$
V = P + g\beta\int_{x_0}^{x}\delta\rho\,d\varphi + \int_{x_0}^{x}p\cos\varepsilon\,de - \int_{x_0}^{x}q\cos\eta\,di ,
$$

(8)
$$
H = Q - \int_{x_0}^{x}p\sin\varepsilon\,de + \int_{x_0}^{x}q\sin\eta\,di
$$

Finally, by differentiating with respect to x and dividing by H, Milankovitch obtains the general equation

(9)
$$
\frac{V}{H} - \frac{dy}{dx} = \frac{1}{H}\left[g\beta\delta\rho\left(\frac{\delta^2}{12\rho} + \xi\right)\sin\varphi\frac{d\varphi}{dx} + p\left(\frac{\delta}{2} + \xi\right)\sin(\varphi - \varepsilon)\frac{de}{dx} + q\left(\frac{\delta}{2} - \xi\right)\sin(\varphi - \eta)\frac{di}{dx}\right].
$$

4) Clear distinction between the direction of the resultant and the tangent to the line of thrust

From the previous equation (9) Milankovitch derives the important conclusion that the direction of the resultant force R at point E does not coincide with the direction of the tangent straight line tt' to the line of thrust at the same point E. As a matter of fact, the ratio $\frac{V}{H} = \tan\psi$ is the trigonometric tangent of the angle ψ between the force R and the x-axis, while the ratio $\frac{dy}{dx} = \tan\alpha$ is the trigonometric tangent of the angle α between the line tt' and the same x-axis. In the general case these angles are different as the second member of the equation (9) is different from zero and Milankovitch observes that at different joints it may be $\psi > \alpha$, $\psi = \alpha$, $\psi < \alpha$, that is the direction of the resultant force may intersect the tangent to the line of thrust.

This result is not completely original. In Milankovitch's analysis, however, it takes a new light and becomes the basis for a critical review of previous studies on the line of thrust where the equation

(10)
$$
\frac{V}{H} - \frac{dy}{dx} = 0
$$

has been wrongly assumed as invariably true.

From the general to the particular: Application of the theory to particular cases of masonry vaults

Starting from the general equation (9) Milankovitch discusses the following particular cases of geometry and load condition.

1) *Die Gewölbe* (the vault)

The term *Gewölbe* is used by Milankovitch when the system is subject only to its own weight and a vertical load at the extrados. Thus, for the *Gewölbe* he takes $q=0$ and $\varepsilon=0$. Moreover he assumes that the joints are perpendicular to the axis of the vault, so that $\overline{\Omega M} = \rho$ represents the curvature radius of the axis at point M and φ is equal to the angle between the tangent line to the axis of the vault and the x-axis. Under these assumptions the equations (9), (7) and (8) become

$$(11) \qquad \frac{V}{H} - \frac{dy}{dx} = \frac{1}{H}\left\{ g\beta\delta\left(\frac{\delta^2}{12\rho} + \xi\right)\sin\varphi\,\frac{d\sigma}{dx} + p\left(\frac{\delta}{2} + \xi\right)\sin\varphi\,\frac{de}{dx} \right\},$$

$$(12) \qquad V = P + g\beta\int_{x_0}^{x}\delta\,d\sigma + \int_{x_0}^{x}p\,de,$$

$$(13) \qquad H = Q.$$

2) *Die Gewölbe gleichen Widerstandes* (the vault of equal resistance)

The vaults of equal resistance fulfil the following two requirements for a given load condition: 1) The equation of the axis of the vault is a particular integral of the differential equation of the line of thrust; 2) The thickness δ of the joints is proportional to the normal component $\overline{N}$ of the resultant force R.

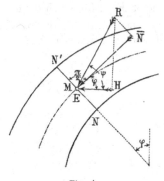

Fig. 4

From the first requirement it follows that the line of thrust may be taken as the one coinciding with the vault axis, that is $\xi = 0$. As a consequence, each joint is subject to constant normal stresses at any point and the following equalities hold (fig. 4)

$$\tan\varphi = \frac{dy}{dx} = \tan\alpha \qquad\qquad \varphi = \alpha \qquad\qquad d\sigma = ds = \frac{dx}{\cos\varphi}\ ,$$

so that equation (11) becomes

(14)
$$\frac{V}{H} - \frac{dy}{dx} = \frac{1}{H}\left\{ \frac{1}{12}g\beta\frac{\delta^3}{\rho}\tan\varphi + \frac{1}{2}p\delta\sin\varphi\frac{de}{dx} \right\}$$

while (12) and (13) do not change.

The second requirement implies $\delta = k\overline{N}$, where k is a constant. As a consequence, the normal stresses are equal at any joint. As $R = \dfrac{H}{\cos\psi}$ and $\overline{N} = H\dfrac{\cos(\psi - \varphi)}{\cos\psi}$, the thickness varies in accordance with the following law

(15)
$$\delta = kH\frac{\cos(\psi - \varphi)}{\cos\psi}\ .$$

3) *Die Stützlinie*

The term *Stützlinie* is used by Milankovitch when both the loads and the joints are vertical, that is when $\varepsilon = 0$, $\eta = 0$ and $\varphi = 0$. Under these assumptions the equation (9) becomes

(16)
$$\frac{V}{H} - \frac{dy}{dx} = 0$$

that is,

(17)
$$\psi = \alpha\ .$$

Thus the *Stützlinie* is the line of thrust for which the direction of the resultant force R coincides with the direction of the tangent tt'. As the loads are assumed to act vertically, a single function $\omega = f(x)$ may be chosen to represent both the self weight and the extrados and intrados load, so that the result is:

(18)
$$V = P + \int_{x_0}^{x}\omega\,dx$$

(19)
$$H = Q$$

and then

(20)
$$\frac{dy}{dx} = \frac{1}{Q}\left\{ P + \int_{x_0}^{x}\omega\,dx \right\}\ .$$

By differentiating with respect to x it follows

(21)
$$\frac{d^2y}{dx^2} = \frac{\omega}{Q}\ ,$$

which is the second order differential equation of the *Stützlinie.*

By integrating the equation (20) and considering that for $x = x_0$ it is $y = y_0$, there follows the explicit equation of the *Stützlinie*:

(22) $\quad y = y_0 + \dfrac{1}{Q} \int\limits_{x_0}^{x} dx \int\limits_{x_0}^{x} \omega\, dx + \dfrac{P}{Q}(x - x_0).$

This equation shows that, where $\omega = 0$, the *Stützlinie* becomes the straight line

(23) $\quad y = y_0 + \dfrac{P}{Q}(x - x_0)$

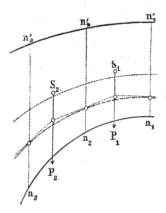

Fig. 5

Then, if the actual load distribution is divided in vertical stripes by the lines $n_1 n_1'$, $n_2 n_2'$... and substituted with the resultant forces P_1, P_2 ... applied at the centres of mass S_1, S_2 ... of each stripe (fig. 5), it follows that the *Stützlinie* becomes a *Stützpolygon* whose vertices lie on the direction of the loads P_1, P_2, ... At the vertical lines $n_1 n_1'$, $n_2 n_2'$, ... the *Stützpolygon* is tangent to the actual *Stützlinie* so that it circumscribes the line of thrust.

4) *Die Kettenlinie*

The term *Kettenlinie* is adopted by Milankovitch for the case of a vault with vanishing thickness, that is $\delta = 0$. In this case the vault becomes a perfectly flexible chain and the equilibrium is possible only if the line of thrust coincides with the chain itself, that is if $\xi = 0$. Thus equation (9) becomes, taking into account that for this limit case it is $de = di = d\sigma = ds$,

(24) $\qquad\qquad \dfrac{V}{H} - \dfrac{dy}{dx} = 0.$

If g_x represents the specific weight of the chain, it follows:

(25) $\qquad\qquad V = P + \int\limits_{x_0}^{x} g_x ds + \int\limits_{x_0}^{x} p \cos \varepsilon\, ds - \int\limits_{x_0}^{x} q \cos \eta\, ds$

$$(26) \qquad H = Q - \int_{x_0}^{x} p\sin\varepsilon\, ds + \int_{x_0}^{x} q\sin\eta\, ds \, .$$

If all the loads act vertically, then the *Stützlinie* coincides with the *Kettenlinie*.

5) Searching for the *mathematische Stärke* (mathematical thickness) in the case of the semicircular arch under its own weight

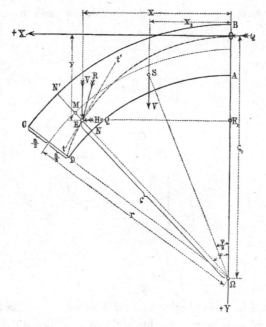

Fig. 6

A detailed application of the general theory is given by Milankovitch for the case of a circular arch of constant thickness $\delta = a$ subject to its own weight (fig. 6). In this case the *Gewölbe* has an axis with constant radius of curvature $\rho = r$.

For the symmetry of the system the resultant force at the crown joint AB has only the horizontal component Q applied at a certain point O with distance ρ_0 from the centre of curvature Ω. Thus, by taking the origin of the reference axes at the same point O, it follows that, for $x = 0$, it is $y = 0$, $V = P = 0$ and $H = Q$.

The equations (11), (12) and (13) for the *Gewölbe* become, taking $p = 0$ and assuming $\beta = 1$ and $g = 1$ for the sake of simplicity,

$$(27) \qquad \frac{V}{H} - \frac{dy}{dx} = \frac{1}{H} ar\left(\frac{1}{12}\frac{a^2}{r} + \xi\right)\sin\varphi\,\frac{d\varphi}{dx} \, ,$$

(28) $$V = ar\int_0^{\varphi} d\varphi = ar\varphi,^6$$

(29) $$H = Q$$

By introducing the polar coordinates $\overline{\Omega E} = \rho$ and $O\hat{\Omega}E = \varphi$, then $\xi = r - \rho$, $y = \rho_0 - \rho\cos\varphi$ and $x = \rho\sin\varphi$, so that

(30) $$dy = -\cos\varphi\, d\rho + \rho\sin\varphi\, d\varphi,$$

(31) $$dx = \sin\varphi\, d\rho + \rho\cos\varphi\, d\varphi.$$

Thus equation (1) becomes

(32) $$\rho(ar\varphi\cos\varphi\, d\varphi + ar\sin\varphi\, d\varphi - Q\sin\varphi\, d\varphi) + (ar\varphi\sin\varphi + Q\cos\varphi)d\rho\frac{1}{12}a(a^2 + 12r^2)\sin\varphi\, d\varphi.$$

Milankovitch shows that this is an exact differential equation. Its integral is

(33) $$\rho(ar\varphi\sin\varphi + Q\cos\varphi) = -\frac{1}{12}a(a^2 + 12r^2)\cos\varphi + C$$

where C is a constant. As for $\varphi = 0$ it is $\rho = \rho_0$, the constant C results

(34) $$C = \rho_0 Q + \frac{1}{12}a(a^2 + 12r^2).$$

Finally, by substituting this value in (33) and observing that $(1 - \cos\varphi) = \sin^2\frac{\varphi}{2}$, Milankovitch finds the explicit equation of the line of thrust

(35) $$\rho = \frac{\rho_0 Q + \frac{1}{6}a(a^2 + 12r^2)\sin^2\frac{\varphi}{2}}{Q\cos\varphi + ar\varphi\sin\varphi},$$

As Milankovitch observes at the end of the first paper, this equation could be directly obtained without deducing the differential equation and then integrating it. This occurs when it is possible to find the analytical expression of the resultant load and its points of application for a finite portion of the system. For instance, if S is the centre of mass of the finite portion $ABNN'$, x_1 its abscissa and V the weight of $ABNN'$, from the rotational equilibrium about E, it follows

$$V(x - x_1) - Qy = 0.$$

For a circular arch of constant thickness $\delta = a$ it is known that

$$\overline{\Omega S} = \frac{1}{6}\frac{a^2 + 12r^2}{r}\frac{\sin\frac{\varphi}{2}}{\varphi}$$

and then

$$x_1 = \overline{\Omega S}\sin\frac{\varphi}{2} = \frac{1}{6}\frac{a^2 + 12r^2}{r}\frac{\sin^2\frac{\varphi}{2}}{\varphi}.$$

Moreover, it is also known that $V = ar\varphi$, $x = \rho\sin\varphi$ and $y = \rho_0 - \rho\cos\varphi$, so that by introducing the previous formulas in the equilibrium equation it follows again (35)

$$\rho = \frac{\rho_0 Q + \frac{1}{6}a(a^2 + 12r^2)\sin^2\frac{\varphi}{2}}{Q\cos\varphi + ar\varphi\sin\varphi}.$$

This explicit equation of the line of thrust still contains two constants to be determined, that is the distance ρ_0 and the horizontal thrust Q. Milankovitch applies it to search for the mathematical thickness of a semicircular arch under its own weight, that is, the value of the thickness corresponding to the unique admissible line of thrust wholly lying within the arch ring, as shown in fig. 7 .

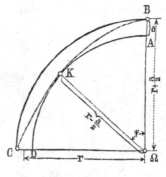

Fig. 7

This is a case already studied in the literature[7] for which it is known that the line of thrust goes through the extrados points B at the crown and C at the springing. Thus the following two boundary conditions must be fulfilled:

(36) $\qquad \rho(0) = r + \dfrac{a}{2}$, which gives $\rho_0 = r + \dfrac{a}{2}$

(37) $\qquad \rho\left(\dfrac{\pi}{2}\right) = r + \dfrac{a}{2}$, which gives $Q = \dfrac{3\pi ar(a+2r) - a(a^2 + 12r^2)}{3a + 12r}$

so that the equation (35) becomes

(38) $\qquad \rho(\varphi) = \dfrac{3\pi r(a+2r)^2 - (a+2r)(a^2 + 12r^2)\cos\varphi}{6\pi r(a+2r)\cos\varphi - 2(a^2 + 12r^2)\cos\varphi + 12r(a+2r)\varphi\sin\varphi}.$

Now, the search for the minimum thickness requires that the line of thrust touches the intrados at a certain point K corresponding to the unknown rupture angle $A\hat{\Omega}K$. This means that the minimum value of the function $\rho(\varphi)$ must be equal to the intrados radius. Thus, where $\dfrac{\partial \rho}{\partial \varphi} = 0$ it must be $\rho = r - \dfrac{a}{2}$.

After some mathematical elaborations Milankovitch finds that the rupture angle is $A\hat{\Omega}K = 54°29'$ [8] and the minimum thickness is $a = 0{,}1075r$.

6) The vault of equal resistance with vertical joints under its own weight

Another special application of the general theory concerns the search for the line of thrust of a vault of equal resistance with vertical joints, subject to its own weight. This case requires that the *Stützlinie* of case 3 fulfills the two requirements of case 2. By taking the origin of the axes at the middle point of the crown joint, for $x = 0$ it must be $y = 0$, $P = 0$, $H = Q$. From the equations (16), (18) and (19) it results

(39) $\dfrac{V}{H} - \dfrac{dy}{dx} = 0$

(40) $V = \displaystyle\int_{x_0}^{x} \delta\, dx$

(41) $H = Q = N$,

so that

(42) $\delta = kN = kQ$

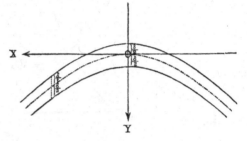

Fig. 8

where k is a constant. By substituting the values of V and H in the differential equation it follows

(43) $k \displaystyle\int_{x_0}^{x} dx = kx = \dfrac{dy}{dx}$.

Then, by integrating and taking into account that for $x = 0$ it is $y = 0$, the result is:

(44) $y = \dfrac{1}{2} kx^2$. [9]

This equation shows that the axis of the vault is a parabola. Then the intrados and extrados lines are also parabolas running at a vertical distance $\pm \delta/2$ from the axis, as shown in fig. 8.

7) Design of the external profile of a retaining wall

Another interesting application of the theory is developed by Milankovitch at the end of his thesis for the design of a retaining wall subject to its own weight, a load K at the top with components P and Q and the lateral pressure of water (fig. 9). Given the internal vertical wall, the purpose is to determine the external profile $BN'C$ in order that the line of thrust is the locus of the external points of the middle third of each horizontal joint, as in this limit case the joints are still subject only to compressive stresses.

By taking the origin of the reference axes at point A and assuming that the line of thrust has equation of the type

(45) $x = f(y),$

then the equation of the external profile $BN'C$ is

(46) $x' = \frac{3}{2} f(y) .$

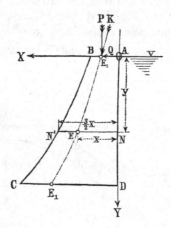

Fig. 9

The general equations (5), (7) and (8) must now be written for the special geometry of the system and the particular load condition under the requirement stated above. This means that $\delta = \frac{3}{2}x$, $\rho = \infty$ (parallel joints), $\rho d\varphi = dy$, $\overline{ME} = \xi = \frac{3}{4}x - x = -\frac{1}{4}x$ (eccentricity of the centre of pressure), $\varphi = \frac{\pi}{2}$ (horizontal joints), $p = 0$ (no extrados load), $q = y$ (internal water pressure, for specific weight equal to one), $\eta = \frac{\pi}{2}$ (horizontal internal load, perpendicular to the internal wall), $di = dy$, $y_0 = 0$. Moreover, by indicating with g the specific weight of masonry and taking the depth $\beta = 1$, the equations (5), (7) and (8) become

(47) $Vdx - Hdy + \frac{3}{8}gx^2 dy = 0$

(48) $V = P + \frac{3}{2}g \int_0^y xdy$

(49) $H = Q + \int_0^y ydy = Q + \frac{1}{2}y^2$

By introducing the formula of V and H and differentiating with respect to x, it follows the differential equation of the line of thrust

(50)
$$\left(8Q - 3gx^2 + 4y^2\right)\frac{d^2y}{dx^2} + 8y\left(\frac{dy}{dx}\right)^2 - 18gx\frac{dy}{dx} = 0 .$$

Milankovitch observes that the general integral of this equation cannot be determined. However, a particular integral can be found for $x_0 = 0$, $P = 0$ and $Q = 0$, that is when the thickness and the load at the top are zero. In this case it results

(51)
$$y = \tfrac{3}{2}\sqrt{g}x ,$$

so that the external profile of the wall is

(52)
$$y = \sqrt{g}x .$$

This equation shows that the wall becomes the right-angled triangle ADC (fig. 10) for which results $\tan C\hat{A}D = 1/\sqrt{g}$.

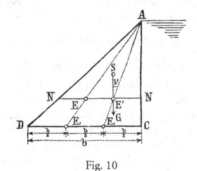

Fig. 10

The line of thrust is the straight line AE_1 , where E_1 is the left point of the middle third at the base DC of the wall. In the absence of the internal pressure of water, the centre of pressure at each joint coincides with the right point of the middle third so that the line of thrust is the straight line AE_1' .

Milankovitch's critical review of previous studies

The previous report shows the mastery of the young Milankovitch in the mathematical treatment of the equilibrium analysis of the arch. On this theoretical basis he also develops some critical remarks that show his mature scientific approach.

To put the problems of applied mechanics in mathematical terms, he says, the scientist has to introduce some hypotheses concerning, for instance, the mechanical behaviour of the materials and the action of the loads. These hypotheses do not fit completely with the reality but they are necessary if the scientist does not want to renounce the use of mathematical tools. Obviously, the assumption of these hypotheses produces a solution whose reliability is directly connected with the reliability of the hypotheses. In this sense this solution is surely approximate but not wrong from a theoretical point of view.

Besides this sort of approximation, a different and more serious type may derive from theoretical mistakes in the *mathematische Stilisierung* of the problem. In particular, Milankovitch recognizes two sources of mistakes invalidating the correctness of a theory: 1) when, by putting the problem in mathematical terms, some circumstances are not correctly taken into account; 2) when the mathematical treatment of a problem correctly modelled is unconsciously wrong.

Having in mind a discussion of some previous contributions to the theory of the line of thrust, Milankovitch observes that the first source of mistakes occurs when the weight dG of the infinitesimal voussoir is supposed to be applied at the middle point M of the joint instead of at the centre of mass G, that is when it is wrongly assumed $\overline{MG} = 0$ instead of $\overline{MG} = \dfrac{\delta^2}{12\rho}$. Now, the condition $\overline{MG} = 0$ holds only if $\rho = \infty$ (case of parallel joints) or if $\delta = 0$ (case of vanishing thickness). Even though the distance $\overline{MG}$ is usually very little, the assumption $\overline{MG} = 0$ cannot be accepted in a good mechanical analysis.

The second source of mistake occurs when the three moments M_g, M_e and M_i are considered as infinitesimals of second order and then neglected in the rotational equilibrium equation. The consequence of this mistake is that the equilibrium equation becomes invariably $\dfrac{V}{H} - \dfrac{dy}{dx} = 0$, so that it would always be $\psi = \alpha$. Now, the condition $\psi = \alpha$, which states the coincidence of the direction of the resultant force R with the tangent straight line tt' to the thrust line, holds only if both the loads and the joints are vertical (case of the *Stützlinie*) or if $\delta = 0$ and the loads acts vertically (case of the *Kettenlinie* under vertical loads). Milankovitch quotes some authors who have erroneously assumed this particular condition even when it is not valid. For instance, Hagen [1846, 1862] takes it for granted in the case of the vault of equal resistance. Similarly, J. Résal [1901] affirms that the coincidence of the line of thrust with the axis of the vault implies the relation $\dfrac{V}{dy} = \dfrac{H}{dx}$. In his deduction of the equation of the line of thrust for an arch subject to its own weight, H.A. Résal [1889, vol. 6, § 238] explicitly writes that the moment M_g is infinitesimal of second order and obtains the wrong result $\dfrac{dy}{dx} = \dfrac{V}{H}$.

Milankovitch also remarks that the first authors who have dealt with the theory of the line of thrust do not committed this sort of mistake. In this sense he quotes Moseley,[10] who clearly states that the direction of the resultant force at a generic joint cuts the line of thrust (called by Moseley line of resistance) and introduces the notion of envelope of the directions of the resultant forces (called by Moseley line of pressure). Milankovitch also cites Dupuit [1870], who gives a correct analytical discussion of the theory of the line of thrust for the special case of a vault under its own weight. In this list we can add also the name of Méry [1840], who gives an analytical proof of the difference between the angular coefficients of the direction of the resultant force and the direction of the tangent to the line of thrust.

Application of the theory to the design of masonry buttresses

In the second paper of 1910 Milankovitch devotes particular attention to the search for the *theoretisch günstigen Formen* of masonry buttresses, that is, for the optimal design under given geometric and static requirements. He considers the general case of a buttress subject to its own weight and a load K applied at a given point E_0 of the top section (fig. 11).

Supposing that the joints are horizontal, as usually occurs for masonry buttresses, and that the line of thrust is the curve E_0EE_1, he follows the reasoning already used for the vault and derives the rotational equilibrium equation of the infinitesimal element $NN'N_1N_1'$ about the point E_1. If V and H are the vertical and horizontal components of the resultant R at the joint NN' and dG is the weight of the voussoir $NN'N_1N_1'$, this equation has the form

(53) $$V\,dx - H\,dy + M_g = 0$$

where dx and dy are the infinitesimal lever arms of the finite forces V and H about E_1, and M_g represents the moment of the infinitesimal weight dG whose finite arm with respect to E_1 may be taken equal to the arm with respect to E because of the infinitesimal distance between E and E_1. Thus the three moments in equation (53) are all infinitesimal quantities of first order.

Starting from equation (53), Milankovitch searches for the optimal shape of the buttress in order to assure that, for given geometry and load condition, the joints are subject only to compressive stresses, as it should be in masonry structures. In particular, he studies three cases of specific interest.

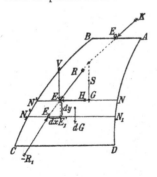

Fig. 11

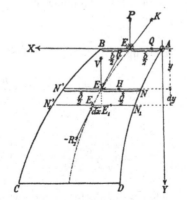

Fig. 12. Profile of a buttress of equal resistance subject to its own weight and a load applied at the top section

1) Design of the buttress of equal resistance

As in the case of the vault, the design of the buttress of equal resistance (fig. 12) fulfils two requirements: 1) the line of thrust must coincide with the locus of the middle points at each horizontal joint; 2) The thickness of the joints must be proportional to the normal component V of the resultant force R.

Thus equation (2) becomes

(54) $\qquad\qquad V dx - H dy = 0$

as the moment M_g is an infinitesimal of second order and may be neglected. Moreover, for the second requirement it must be

(55) $\qquad\qquad \delta = kV$

where k is a constant.

If P and Q are the vertical and horizontal components of the force K applied at the middle point E_0 of the top section, it results, taking into account that Q is the only horizontal force on the buttress and recalling equation (55),

(56) $\qquad\qquad H = Q$

(57) $\qquad\qquad V = \dfrac{\delta}{k} = P + g \int_0^y \delta \, dy$

where g is the specific weight of the masonry.

Differentiating equation (5) with respect to y and separating the variables, it follows that

(58) $\qquad\qquad \dfrac{d\delta}{\delta} = kg dy$.

Thus, by integrating under the boundary condition that for $y = 0$ it is $\delta = b$, the following law for the thickness is obtained

(59) $\qquad\qquad \delta = b e^{kgy}$.

Now from equations (54), (56), (57) and (59) it comes

(60) $\qquad\qquad Q \dfrac{dy}{dx} = P + gb \int_0^y e^{kgy} \, dy = P + \dfrac{b}{k} e^{kgy}$,

so that

(61) $\qquad\qquad x = kQ \int \dfrac{dy}{kP + b e^{kgy}} + C$.

Milankovitch shows that the function under the integral sign may be put in the form

(62) $\qquad \dfrac{1}{kP + b e^{kgy}} = \dfrac{1}{kP} \dfrac{kP + b e^{kgy} - b e^{kgy}}{kP + b e^{kgy}} = \dfrac{1}{kP} \left\{ 1 - \dfrac{b e^{kgy}}{kP + b e^{kgy}} \right\} = \dfrac{1}{k^2 gP} \left\{ kg - \dfrac{kgb e^{kgy}}{kP + b e^{kgy}} \right\}$

and obtains

(63) $$x = \frac{Q}{kgP}\left\{kgy - \log_{nat}\left(kP + be^{kgy}\right)\right\} + C$$

As for $y = 0$ it must be $x = \frac{b}{2}$, the constant C becomes

$$C = \frac{b}{2} + \frac{Q}{kgP}\log_{nat}\left(kP + b\right)$$

so that the middle line of the buttress of equal resistance, which coincides with the line of thrust, has the equation

(64) $$x = \frac{b}{2} + \frac{Q}{kgP}\left\{kgy - \log_{nat}\frac{kP + be^{kgy}}{kP - b}\right\}.$$

In the particular case for which $Q = 0$, from equation (64) it follows that the middle line becomes the vertical line $x = \frac{b}{2}$, while the law of the thickness is always given by (59).

2) Design of a buttress with vertical internal line, when the line of thrust goes through the middle point of the joints

A second application concerns the design of a buttress with vertical internal line, under the requirement that the line of thrust coincide with the locus of the middle points of the horizontal joints (fig. 13).

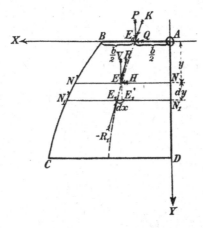

Fig. 13. Profile of a buttress with vertical internal line, when the line
of thrust coincides with the axis of the buttress

Also in this case the rotational equilibrium equation results

(65) $$V dx - Q dy = 0$$

as the term M_g is infinitesimal of second order. The vertical component V at the generic joint is

(66) $$V = P + 2g \int_0^y x \, dy$$

so that

(67) $$\frac{dy}{dx} = \frac{V}{Q} = \frac{P}{Q} + \frac{2g}{Q} \int_0^y x \, dy \, .$$

Differentiating with respect to x it follows

(68) $$\frac{d^2 y}{dx^2} = \frac{2g}{Q} x \frac{dy}{dx}$$

Thus, by integrating Milankovitch finds

(69) $$\frac{dy}{dx} = Ce^{\frac{g}{Q}x^2}$$

where C is a constant to be determined. Now, as the equation $\frac{dy}{dx} = \frac{V}{Q}$ states that the resultant force at each joint is tangent to the line of thrust, it must be for the top section:

(70) $$\left.\frac{dy}{dx}\right\}_{x=\frac{b}{2}; \, y=0} = \frac{P}{Q} = Ce^{\frac{g \, b^2}{Q \, 4}} \, ,$$

so that

(71) $$C = \frac{P}{Q} e^{-\frac{g \, b^2}{Q \, 4}}$$

Thus equation (69) becomes

(72) $$\frac{dy}{dx} = \frac{P}{Q} e^{-\frac{g \, b^2}{Q \, 4}} e^{\frac{g}{Q}x^2}$$

and its integral results

(73) $$y = \frac{P}{Q} e^{\frac{g \, b^2}{Q \, 4}} e^{\frac{g}{Q}x^2} \int e^{\frac{g}{Q}x^2} \, dx$$

This is the equation of the line of thrust. To calculate the integral contained in this equation Milankovitch refers to the method suggested by Stieltjes [1886].

---

3) Design of a buttress with a vertical internal line, when the line of thrust goes through the external points of the middle third

A third application deals with the design of a buttress with a vertical internal line, under the requirement that the line of thrust coincide with the locus of the external points of the middle third of the horizontal joints (fig. 14).

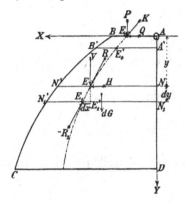

Fig. 14. Profile of a buttress with a vertical internal line, when the line of thrust goes through the external points of the middle third

In this case the term M_g is infinitesimal of first order as the moment of the weight dG has a finite lever arm with respect to E_1, and this arm may be taken equal to the distance of dG from E, that is $\dfrac{x}{4}$.

Thus the moment M_g is

(74)
$$M_g = \frac{x}{4} dG = \frac{x}{4}\frac{3}{2} xgdy = \frac{3}{8} gx^2 dy \ ,$$

and the rotational equilibrium equation results

(75)
$$V dx - Q dy + \frac{3}{8} gx^2 dy = 0 \ .$$

The vertical component V at the generic joint is

(76)
$$V = P + \frac{3}{2} g \int_0^x x dy \ ,$$

so that the equilibrium equation becomes

(77)
$$P + \frac{3}{2} g \int_0^x x dy - Q \frac{dy}{dx} + \frac{3}{8} gx^2 \frac{dy}{dx} = 0 \ .$$

Differentiating with respect to x it follows

(78)
$$\left(\frac{3}{8}gx^2 - Q\right)\frac{d^2y}{dx^2} + \frac{9}{4}gx\frac{dy}{dx} = 0$$

and then

(79)
$$\frac{\dfrac{d^2y}{dx^2}dx}{\dfrac{dy}{dx}} = -\frac{18gx}{3gx^2 - 8Q}dx .$$

Now, by integrating Milankovitch finds

(80)
$$\frac{dy}{dx} = C\left(3gx^2 - 8Q\right)^{-3}$$

The equation (75) for the rotational equilibrium of the infinitesimal voussoir about E'_0 gives

(81)
$$Pdx - Qdy + \frac{3}{8}gx_0^2\,dy = 0 .$$

Thus, taking into account (80), for the top section it must be

(82)
$$\left.\frac{dy}{dx}\right\}_{x=x_0;y=0} = C\left(3gx_0^2 - 8Q\right)^{-3} = -\frac{8P}{3gx_0^2 - 8Q}$$

so that

(83)
$$C = -8P\left(3gx_0^2 - 8Q\right)^2 .$$

By replacing the value of C equation (80) becomes

(84)
$$\frac{dy}{dx} = -8P\left(3gx_0^2 - 8Q\right)^2\left(3gx^2 - 8Q\right)^{-3}$$

and its integral results, by putting $\dfrac{8Q}{3g} = k^2$,

(85)
$$y = -\frac{8}{3}\frac{P}{g}\left(x_0^2 - k^2\right)^2 \int \frac{dx}{\left(x^2 - k^2\right)^3} .$$

Milankovitch observes that the function under the integral sign can be written as the following sum:

(86)
$$\frac{1}{\left(x^2 - k^2\right)^3} = \frac{1}{8k^3}\frac{1}{(x-k)^3} - \frac{3}{16k^4}\frac{1}{(x-k)^2} + \frac{3}{16k^5}\frac{1}{x-k} - \frac{1}{8k^3}\frac{1}{(x+k)^3} - \frac{3}{16k^4}\frac{1}{(x+k)^2} - \frac{3}{16k^5}\frac{1}{x+k} .$$

Thus, taking into account that if $x = x_0$, then $y = 0$, Milankovitch obtains the equation of the line of thrust $E_0 EE_1 E_1'$

(87) $\qquad y = \dfrac{2}{3}\dfrac{P}{k^2 g}\left(x_0^2 - k^2\right)^2\left\{\dfrac{\dfrac{x}{\left(x^2 - k^2\right)^2} - \dfrac{x_0}{\left(x_0^2 - k^2\right)^2} - \dfrac{3}{2k^2}\left[\dfrac{x}{x^2 - k^2} - \dfrac{x_0}{x_0^2 - k^2}\right] -}{-\dfrac{3}{4k^3}\log_{nat}\dfrac{(x-k)(x_0+k)}{(x+k)(x_0-k)}}\right\}$

If $P = 0$, that is if at the top section acts only the horizontal component Q (fig.15), then the function y is defined only for $x_0 = k = \sqrt{\dfrac{8Q}{3g}}$. In this case the line of thrust becomes the vertical line with equation $x = \overline{AE_0} = \sqrt{\dfrac{8Q}{3g}}$ and consequently the external profile BC is also vertical and has equation $x = \overline{AB} = \dfrac{3}{2}\overline{AE_0} = \sqrt{\dfrac{6Q}{g}}$.

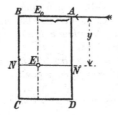

Fig. 15

As Milankovitch observes, a remarkable property of this case is that the resultant force at the top section is perpendicular to the line of thrust.

Notes

1. [Milankovitch 1907] reproduces the content of the doctoral thesis, with few revisions and some different notations. [Milankovitch 1910] is an application of the general theory of the thrust line for the optimal design of masonry buttresses.
2. The rediscovery of this mechanical tradition starts with the first historical investigation by Jacques Heyman [1972]. Significant historical studies on the theory of the arch have been given by Edoardo Benvenuto [1981, 1991]. Special attention to the role of geometry and equilibrium in the history of masonry vaulted structures is given by Santiago Huerta [2001, 2004]. A historical analysis of the pre-elastic methods is contained in Federico Foce [2002, 2005].
3. As far as we know, the first discussion of Milankovitch's two published papers appears in [Huerta 1990]. Another discussion has been given by the author of the present paper [Ageno, et. al, 2004].
4. For more details on Milankovitch (b. 28 May 1879 in Dalj near Osijek, (Austria-Hungary) – d. 12 December 1958 in Belgrade), see the many websites on his scientific work.
5. [Milankovitch 1904, 2; 1907, 3]. Obviously, the centre of mass G of the infinitesimal voussoir does not lie on the joint line and in this sense the term *Schwerpunkt der Fuge* is rather improper. However, as we shall see in the following discussion where the weight dG of the voussoir is correctly applied at the true centre of mass S (see fig. 3), this choice is taken by

Milankovitch just to point out that G, projection of S on the joint line, has the same finite distance from the middle point M.

6. In the first printed paper [1907a], but not in the manuscript of the thesis, this formula is erroneously written $V = ar \int_0^\varphi d\varphi = \mathrm{arc}\varphi$.

7. On this point Milankovitch quotes Ritter [1899] and Pilgrim [1877].

8. Milankovitch adds that for the rupture angle Ritter has given 54°10′ and Pilgrim 54°27′. Other authors, not quoted by Milankovitch, have obtained quantitative results on the minimum thickness of a semicircular arch under its own weight, even though by different methods of analysis. For instance: Couplet, who for the minimum thickness gives 0.1061 of the intrados radius by taking the a priori value of 45° for the rupture angle [Couplet 1732]; Monasterio, who for the thickness gives the range 0.111-0.125 of the intrados radius and for the rupture angle the range 54-56° [Monasterio, ca.1800]; Petit, who obtains the good approximation 0.114 of the intrados radius with rupture angle at 54° [Petit 1835].

9. Milankovitch writes $y = kx^2$, perhaps including the factor ½ in the constant k.

10. Milankovitch refers to the German translation by Scheffler of [Moseley 1845].

References

AGENO, A., A. BERNABÒ, F. FOCE, A. SINOPOLI. 2004. Theory and history of the thrust line for masonry arches. A Brief Account. Pp. 1-10 in *Arch Bridges IV. Advances in assessment, structural design and construction*, P. Roca and C. Molins (eds.) Barcelona: CIMNE.

BENVENUTO, E. 1981. *La scienza delle costruzioni e il suo sviluppo storico*. Florence: Sansoni (rpt. Rome: Edizioni di Storia e Letteratura, 2006).

———. 1991. *An introduction to the history of structural mechanics*. Berlin: Springer.

COUPLET, P. 1732. Seconde partie de l'éxamen de la poussée des voûtes. *Mémoires de l'Académie Royale des Sciences*, 1730 : 117-141.

DUPUIT, J. 1870. *Traité de l'équilibre des voûtes et de la construction des ponts en maçonnerie*. Paris: Dunod.

FOCE, F. 2002. Sulla teoria dell'arco murario. Una rilettura storico-critica. Pp. 129-213 in A. Becchi and F. Foce, *Degli archi e delle volte. Arte del costruire tra meccanica e stereotomia*. Venice: Marsilio.

———. 2005. On the safety of the masonry arch. Different formulations from the history of structural mechanics. Pp.117-142 in *Essays in the history of theory of structures*, S. Huerta (ed.). Madrid: Instituto Juan de Herrera.

HAGEN, G.. 1846. Über die Form und Stärke der gewölbten Bogen. *Abhandlungen der königlichen Akademie der Wissenschaften zu Berlin*, Jahr 1844: 51-72

———. 1862. *Über Form und Stärke gewölbter Bogen und Kuppeln*. Berlin: Ernst und Korn.

HEYMAN, J. 1966. The stone skeleton. *International Journal of solids and structures* 2: 249-279.

———. 1972. *Coulomb's memoir on statics. An essay in the history of civil engineering*. Cambridge: Cambridge University Press.

HUERTA, S. 1990. Diseño estructural de arcos, bóvedas y cúpulas en España, ca. 1500-1800, Ph.D. thesis, E.T.S. de Arquitectura de Madrid.

———. 2001. Mechanics of masonry vaults: the equilibrium approach. Pp. 47-69 in *Historical Constructions 2001*, Lourenço P. B., Roca P. (eds.). Guimaraes: Universidade do Minho.

———. 2004. *Arcos, bóvedas y cúpulas. Geometría y equilibrio en el cálculo tradicional de estructuras de fábrica*. Madrid: Instituto Juan de Herrera, 2004.

MÉRY, E. 1840. Sur l'équilibre des voûtes en berceau. *Annales des Ponts et Chaussées* 19 (I sem.): 50-70.

MILANKOVITCH, M. 1904. Beitrag zur Theorie der Druckkurven. Dissertation zur Erlangung der Doktorwürde, K.K. technische Hochschule, Vienna.

———. 1905. *Beitrag zur Theorie der Betoneisenträger*. Vienna.

————. 1907a. Theorie der Druckkurven. *Zeitschrift für Mathematik und Physik* **55**: 1-27.

————. 1907b. *Die vorteilhafteste Konstruktionshöhe und Verlagsweite der Rippen der Hennebiqueschen Decke*. Vienna.

————. 1908. *Eisenbetondecke mit Isoliereinlagen System Milankovitch-Kreutz*. Vienna.

————. 1910. Zur Statik der massiven Widerlager. *Zeitschrift für Mathematik und Physik* **58**: 120-128.

MONASTERIO, J. No date (ca. 1800). Nueva teórica sobre el empuje de bóvedas. Unpublished manuscript. Library of the E.T.S. de Ingenieros de Caminos, Canales y Puertos, Universidade Politécnica de Madrid.

MOSELEY, H. 1845. *Die mechanischen Prinzipien der Ingenieurkunst und Architektur*. German translation by H. Scheffler. Braunschweig.

PETIT. 1835. Mémoire sur le calcul des voûtes circulaires. *Mémorial de l'Officier du Génie* **12** : 73-150.

PILGRIM, L. 1877. *Theorie der kreisförmiger symmetrischen Tonnengewölbe von konstanter Dicke, die nur ihr eigenes Gewicht tragen*. Stuttgart: K. Wittwer.

RÉSAL, J. 1901. *Stabilité des constructions*. Paris : Béranger.

RÉSAL, H. A. 1873-1889. *Traité de mécanique générale*. Paris: Gauthier-Villars.

RITTER, A. 1899. *Lehrbuch der Ingenieurmechanik*. Leipzig.

STIELTJES, T.J. 1886. Sur les racines de l'équation $X_n = 0$. *Acta mathematica* **9**: 167-176.

About the author

Federico Foce is a researcher in Structural Mechanics in the Department of Sciences of Architecture at the University of Genoa. His studies focus on the mechanics of solids and structures, with a particular interest in the historical development of scientific theories applied to construction of buildings. He has undertaken historical studies on the theory of the elasticity of solids and on the theory of vaulted masonry structures, and is co-author with Antonio Becchi of *Degli archi e delle volte. Arte del costruire tra meccanica e stereotomia* (Venice: Marsilio, 2002). As co-founder of the Edoardo Benvenuto Association for Research on the Science and Art of Building in their Historical Development, he has organised international conferences and has co-edited (with Antonio Becchi, Massimo Corradi, and Orietta Pedemonte) the book series entitled *Between Mechanics and Architecture*.

Santiago Huerta

E. T.S. de Arquitectura
Universidad Politécnica de Madrid
Avda. Juan de Herrera, 4
28040 Madrid SPAIN
Santiago.Huerta@upm.es

Keywords: oval domes, history of
engineering, history of
construction, structural design

Research

Oval Domes: History, Geometry and Mechanics

Abstract. An oval dome may be defined as a dome whose plan or profile (or both) has an oval form. The word "oval" comes from the Latin *ovum*, egg. The present paper contains an outline of the origin and application of the oval in historical architecture; a discussion of the spatial geometry of oval domes, that is, the different methods employed to lay them out; a brief exposition of the mechanics of oval arches and domes; and a final discussion of the role of geometry in oval arch and dome design.

Introduction

An oval dome may be defined as a dome whose plan or profile (or both) has an oval form. The word Aoval@ comes from the Latin *ovum*, egg. Thus, an oval dome is egg-shaped. The first buildings with oval plans were built without a predetermined form, just trying to enclose a space in the most economical form. Eventually, the geometry was defined by using circular arcs with common tangents at the points of change of curvature. Later the oval acquired a more regular form with two axes of symmetry. Therefore, an "oval" may be defined as an egg-shaped form, doubly symmetric, constructed with circular arcs; an oval needs a minimum of four centres, but it is possible also to build ovals with multiple centres.

The preceding definition corresponds with the origin and the use of oval forms in building and may be applied without problem up to, say, the eighteenth century. From that point on, the study of conics in elementary courses of geometry taught the learned people to consider the oval as an approximation of the ellipse, an "imperfect ellipse": an oval was, then, a curve formed from circular arcs which approximates the ellipse of the same axes. As we shall see, the ellipse has very rarely been used in building.

Finally, in modern geometrical textbooks an oval is defined as a smooth closed convex curve, a more general definition which embraces the two previous, but which is of no particular use in the study of the employment of oval forms in building.

The present paper contains the following parts: 1) an outline of the origin and application of the oval in historical architecture; 2) a discussion of the spatial geometry of oval domes, i.e., the different methods employed to lay them out; 3) a brief exposition of the mechanics of oval arches and domes; and 4) a final discussion of the role of geometry in oval arch and dome design.

Historical outline of the origin and application of the oval in historical architecture

The first civilizations: Mesopotamia and Egypt

Rounded forms, many times not geometrically defined, were used in building from the most remote antiquity. These rounded forms may be called "oval". What the ancient builders were looking for was the most simple and economical way to enclose a space. As techniques were perfected, some of these plans were geometrically defined using cords and pegs to control their contours, i.e., employing circular arcs or combinations of them.

These enclosures were first covered by masonry in about 4000 B.C., by cantilevering the stones forming successive rings, until the space is closed at the top. This is what we call now a "false dome". Fig. 1 shows the most ancient remains discovered so far in Asia Minor. Domes were used to form "stone huts" and the technique was developed, no doubt, in the context of permanent settlements associated with agriculture. It is a form of what we today call vernacular construction. The same technique of building has survived in some countries until the present day. (In Spain, for example, the vernacular buildings of *piedra seca*, dry stone, in Mallorca are similar to those first examples in Asia Minor [Rubió 1914].)

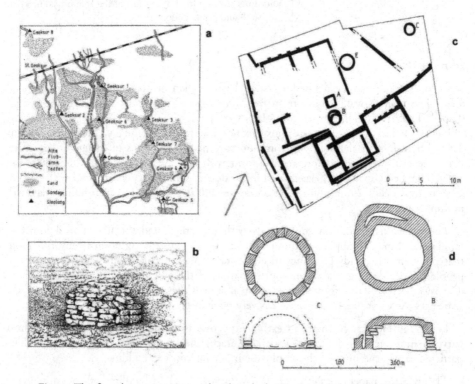

Fig. 1. The first domes covering a closed oval plan in Asia Minor ca. 4,000 B.C. [Baitimova 2002]

The invention of the arch apparently came later than that of the dome. The first arches were built in Mesopotamia or Egypt circa 3500 B.C. to construct the permanent covering of tombs. The books of Besenval [1984] and El-Naggar [1999] contain the most information on arch and vault building in those times.

The first arches were built with crude bricks. It was discovered that if the bricks are disposed in a certain manner in space they remain stable, their weight being transferred from one brick to the next until reaching the earth: the same force which tries to drag the bricks to the earth keeps them in position. It was an amazing invention and an enormous step forward from the more common custom of simply piling the bricks to form walls. (The practice of brick wall building was itself an invention which evolved very slowly before the bricks and the different bondings were developed; see Sauvage [1998].)

In the first two millennia the builders experimented with several types of arches and vaults and there is no direct line of progress towards the voussoir arch with radial joints, which is our conceptual model. A perusal of the hundreds of surveys contained in the books of Besenval and El-Naggar makes evident a long period of "eclectic" experimentation, in which several forms and types of arches co-existed. Among them appeared the first oval vaults (fig. 2). Some of those vaults were built without centring, by building successive flat slices against a wall where the form of the arch was first drawn. The technique is still used in the north of Africa [Fathy 1976] (fig. 3).

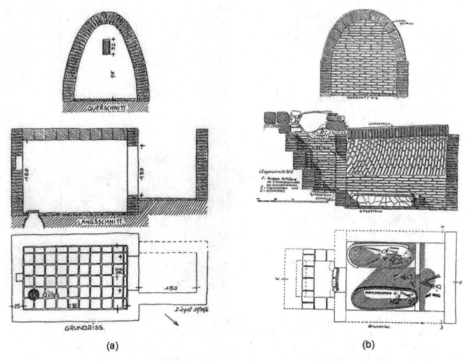

(a)   (b)

Fig. 2. Oval arches in Asur, Mesopotamia. a) With radial centres; b) Arches built without centring, by constructing successive slices leaned one against the next [Besenval 1984]

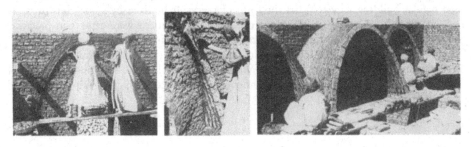

Fig. 3. Building oval barrel vaults in North Africa in the 1970's [Fathy 1976]

The first vaults were quite small, with spans of only about one or two meters, just enough to cover the tomb. This size favoured experimentation: the vault, if not of an adequate form, will distort and the observation of the movements gave the builders a "feeling" for the more adequate forms. Fig. 4a shows one of the plates of the book of El-Naggar [1999] which explains clearly the kind of forms adopted for the vaults. To an architect or engineer with some experience in masonry structures it will be evident that the vault at the bottom right side is the safest, adopting an oval form which will amply contain the trajectory of compressions (the line of thrust or inverted catenary) within the arch. Choisy [1904] was the first to point this fact as the origin of the oval arches (fig. 5a). (On the design of masonry arches see [Heyman 1995] and [Huerta 2006].)

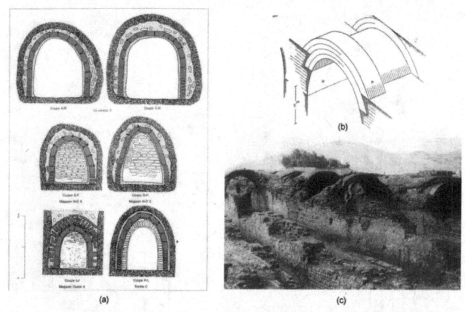

(a)   (b)   (c)

Fig. 4. a) Different Egyptian oval vaults, showing different degrees of distortion from the oval form [El-Naggar 1999]. The bottom-left vault, which shows no distortion, presents no danger of collapse; b) Construction of the vaults of the Ramesseum [Choisy 1904]; c) Barns of the Ramesseum [El-Naggar 1999]

When vaults grew bigger in the second millennium B.C. – for example, the vaults covering the barns of the Ramesseum (fig. 4b and c) have spans of almost five meters – a good regular building required that the form of the vault be fixed by some construction. The oval forms had to be defined geometrically. The Egyptians were experts in practical geometry using pegs and strings, and a form composed of circular arcs is the most logical. Choisy [1904], observing the form of many vaults, considering the practical geometry of the Egyptians (the 3-4-5 right triangle), and applying the logic of building, proposed a simple oval form and suggested how it might be constructed using a simple system of strings (fig. 5).

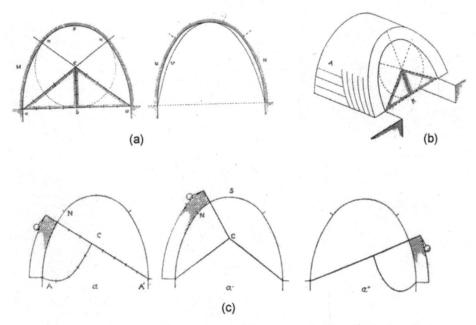

Fig. 5. The geometry and construction of Egyptian ovals, after Choisy [1904]. a) Egyptian oval with the 3-4-5 triangle and comparison with the catenary; b) Possible employment of the oval with leaning brick slice-arches; c) Use of a string method to draw the oval

A geometrical construction is not the only possibility. The mason may sketch the profile of the vault on a wall, perhaps making several corrections until he is satisfied with a certain form. Then, he may fix the form my drawing an horizontal line and measuring the vertical distances to it. Indeed, this was the method followed in a diagram from the Third Dynasty (3000-2700 B.C.) near the Step Pyramid of Saqqara [Gunn 1926] (fig. 6a and b). If the separation between the vertical lines is considered to be uniform, the profile does not correspond to the preceding "typical" oval or a circular arch, and this supports the previous hypothesis. However, Daressy [1927] demonstrated that if the last interval is presumed to be shorter than the others a circular arc may be somewhat adjusted to the curve (fig. 6c). The present author has adjusted an oval, following a simple geometrical construction (fig. 6d). Many other curves may be tried, but any interpretation should be made with caution, considering the historical context and the logic of practical building at that time.

Some scholars claim to have found ellipses and not ovals in the form of the Egyptian arches. In particular, the French archaeologist Daressy [1907] attributed an elliptical form to a drawing of the profile of an arch corresponding to the vaults of one of the tombs of Ramses VI (twelfth century B.C.). This hypothesis has been accepted as true by many scholars who have discussed the laying out of Egyptian arches. (See, for example, [Arnold 1991] and [Rossi 2004]). Heisel [1991] hinted at the possibility of employing a bent wooden strip to lay out the curve. ([Cejka 1978] has explored this method in his research on Islamic arches).

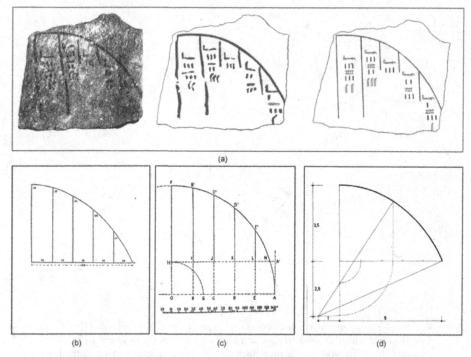

(a)

(b)　　　　　(c)　　　　　(d)

Fig. 6. a) The oldest diagram of an arch, Third Dynasty, 3000-2700 B.C. (photo and drawings after [Gunn 1926]; b) Drawing by Gunn [1926]; c) Geometrical interpretation as an arc of circle by Daressy [1927]; d) Oval which approximates to Gunn=s drawing (author)

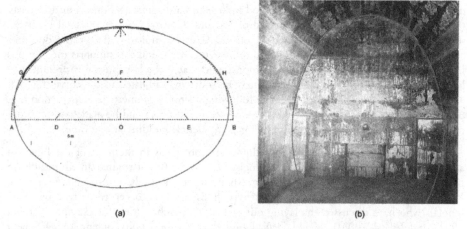

(a)　　　　　　　　　(b)

Fig. 7. a) Layout of an Egyptian oval dome. Comparative drawing of an Egyptian oval with an ellipse with foci in D and E ([El-Naggar 1999] after [Daressy 1907]). Choisy=s oval has been superimposed by the author in a thin continuous line; b) Stone vault of an Egyptian tomb of oval profile [El-Naggar 1999]. Choisy=s oval superimposed by the author

I disagree strongly with the hypothesis that the Egyptians used elliptical forms. The ellipse is a mathematical curve which was defined by Greek mathematicians the fourth century B.C. [Heath 1981]. The supposition that they have discovered by chance the string method for laying out the ellipse (the so-called "gardener=s method") is also quite difficult to accept, given the long and painful birth of even the most simple and "obvious" inventions. In fact, Choisy=s Egyptian oval adapts itself as well or better than the ellipse to the diagram, as fig. 8a makes evident, and if any geometrical construction was used, this appears a much more probable hypothesis. Finally, a drawing found in a wall of Luxor's Temple (figs. 8a and b) [Borchardt 1896] provides evidence of the employment of the oval by the Egyptians. In this case it is evident that the form corresponds to an oval, because of the great difference of the two radii. This is a very strong argument, to be added to the others, against the "ellipse hypothesis".

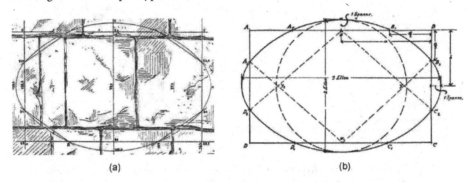

(a)   (b)

Fig. 8. Drawing of an oval discovered by Borchartd in 1896 in a wall of the Temple of Luxor. a) Borchardt's reproduction of the original drawing; b) One of his hypothesis for the geometrical generation of the oval [Borchartd 1896]

Greece and Rome

The Greeks knew the arch. Since stone was the usual building material, they employed the voussoir arch, almost without exception with a semicircular form. The arch was employed in secondary buildings, in sewers or for the gates of the city walls [Boyd 1978, Dornisch 1992]. Some cantilevered domes approached the oval form, but there was no systematic use of the ovals as in the brick architecture of Mesopotamia or Egypt.

The oval form appeared in Europe in Roman times for the design of amphitheatres (see Wilson Jones 1993]) (fig. 9a). Again, some scholars have tried to prove the use of ellipses instead of ovals. However, the oval form is the natural form for laying out the stands of the amphitheatre: it is impossible to construct parallel ellipses, and the only logical method is to use oval forms made of circular arcs. In any case, the differences between ellipses and ovals for the usual proportions are so small that, in fact, one see what one wants to see. Even the most precise mensuration does not serve to settle the matter [Rosin 2005]. It is not a matter of mensuration, but of the history of building traditions.

It appears that the Romans did not built oval domes: the central symmetry was considered a requisite and even in the experimentation of the domes in Hadrian's Villa all the forms present a centralized character [Rasch 1985]. Some exceptions may be found in the apses of thermae [Lotz 1955], and it is usually presumed that the octagon of the church of St. Gereon in Cologne rests on the oval foundations of a previous Roman building [Götz

1968] (fig. 9b) and [Krautheimer 1984] (fig. 9c). Choisy [1873] discovered that in the bridge of Narni, the central span of the inclined road was adapted by using an arch formed from two quarter-circles of different radii. Perhaps more examples can be found, but it appears that the Romans used the oval form for arches and domes only in exceptional cases.

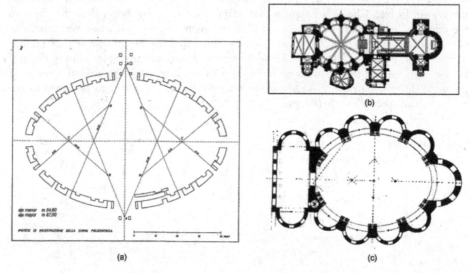

(a)   (b)   (c)

Fig. 9. a) One of the hypothesis concerning the oval geometry of the Roman Coliseum [1993]; b) Medieval octagon of the church of St. Gereon in Cologne; it is supposed to be built on the foundations of an oval Roman building [Götz 1968]; c) reconstruction by Krautheimer [1984]

The Middle Ages

Most medieval domes have a centralised form, probably due to the Roman influence. In Spain most Romanesque churches pertain to this type. However, there are also some exceptions, and the church of Santo Tomás de Olla, dated by Gómez Moreno [1919] to the tenth century, presents an octagonal dome on an oval plant, 6 x 5.5 m, surrounded by horseshoe arches (Fig, 10).

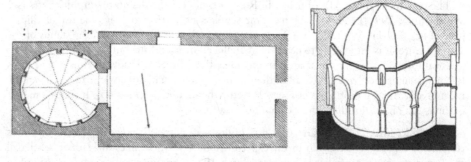

Fig. 10. Church of Santo Tomás de Olla, Spain, built in the tenth century [Gómez Moreno 1919]

In France, Chappuis [1976] has made what appears to be the only exhaustive study of the use of the oval form in the Middle Ages. He has studied some 400 Romanesque churches in the south of France which present some kind of oval arches or domes. Of these, 130 have domes with an oval plan. Fig. 11a shows Chappuis's classification of Romanesque ovals. Fig. 11b shows clearly the oval plan in two churches. Chappuis has found, then, a precedent in Europe for the Renaissance oval domes, which so far has not been noted by historians of architecture.

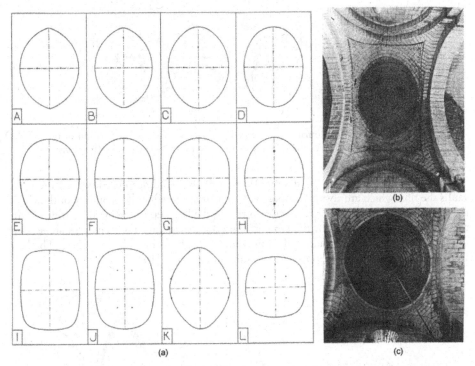

(a) (b) (c)

Fig. 11. The oval form in French Romanesque churches [Chappuis 1976]. a) Classification of ovals; b) Two examples of churches with oval plans: above, Saint-Martin de Gurçon, Dordogne, and below, Balzac, Charente

Another element in which oval arches can be found in Romanesque churches are the groined vaults, resulting from the intersection of two barrel vaults. It may that the medieval masons knew some method to lay out approximately the curve of the groin before construction, and the technique of the "lengthened arch" which will be discussed below, could have been used. But there are other techniques which make possible the construction without the physical building of the diagonal centring, as may be seen in the drawing by Mohrmann [Ungewitter 1890] (fig. 12). (The Romans may have used the technique in building their groined vaults.)

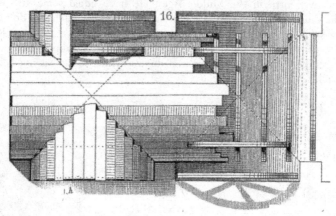

Fig. 12. Building of a Romanesque groined vault without knowing the form of the groin. The intersection is made "physically" with the wooden planks [Ungewitter 1890]

The Gothic is based on a great simplification of the building procedures. In the cross vaults the ribs, which are always composed of circular arcs, define the geometry of the vault, the masonry shell closing the space between ribs in the last step of the building (fig. 13). In the simple quadripartite vault, the cross ribs are semicircles and the transverse arches and the formerets (wall ribs) are adjusted to the desired height using pointed arches. (The best study on the geometry of the Gothic ribs is still that of Willis [1843].)

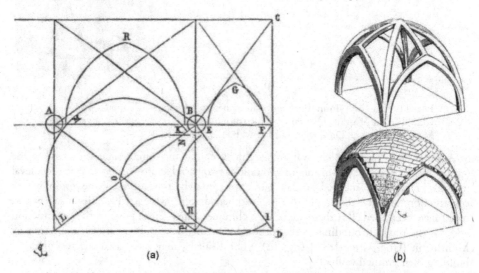

(a)   (b)

Fig. 13. The geometry of the Gothic cross vault is defined by that of the ribs, which are made of circular arcs. a) The technique permits correspondence with the forms of adjacent vaults; b) The ribs are built before and then the shell is closed [Viollet-le-Duc 1858]

The idea of defining beforehand the curve of intersection between two barrel vaults – difficult to lay out and construct – by a simple circular arc, seems to come from Byzantium; Choisy [1883] studied the matter in depth. The Byzantines did not use ribs to reinforce the groins, but this, in essence, is the only conceptual difference between the laying out of Byzantine and Gothic cross vaults (fig. 14).

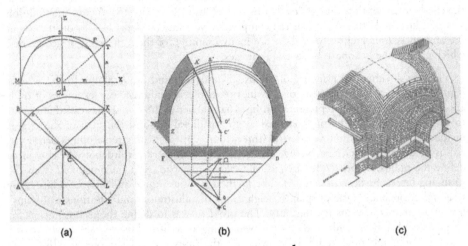

(a)　　　　　　　　　(b)　　　　　　　　　(c)

Fig. 14. Laying out of the groins in a Byzantine cross vault. a) The two groins and the four perimeter arches are circular segments, defined beforehand; b) The shell is "generated", hypothetically, by a system of strings rotating around an axis [Choisy 1883]; c) The different arch-slices are built without centring until the vault is closed [Ward-Perkins 1958]

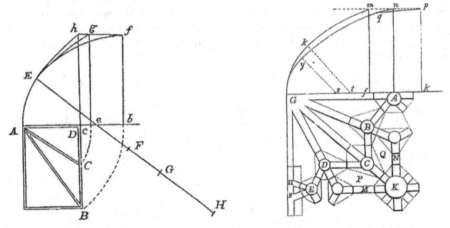

Fig. 15. Use of oval segmental ribs to define the geometry of late Gothic English vaults [Willis 1843]

Gothic geometry is coherent with the character of Gothic building: a requisite of economy, in the broadest sense of the term, permeates the whole Gothic building. The ingenuity of the master masons in solving easily, by simple methods, the most complicated problems still fascinate the historians of Gothic architecture today. In high Gothic, the ribs were always either semicircular or pointed arches of two centres. The flexibility of the system has been best explained in the drawings by Viollet-le-Duc (fig. 13a, above).

In late Gothic, however, in England, Germany and Spain, the Gothic masters began to employ oval forms, generated by the tangency of circles of different radii. The first to note the employment of this kind of "oval" ribs was Willis [1843] (fig. 15). Again, the tradition was well known in the Orient. In the comprehensive study of Cejka [1978] on the use of the arch in Islamic architecture, there are many examples of the same kind of geometries, which predate the European by several centuries.

The first documented evidence of the use of ovals is found in Germany. In the Wiener Sammlung, the richest collection of drawings of medieval buildings, two examples of ovals are found; one of them is represented in fig. 16a. Bucher [1968, 1972] has called attention to them and describes the method used by the medieval masons: the span of the arch is divided into a certain number of parts (three, four, ...); the two extreme points define the radius of the smaller circles; finally, on the axis of symmetry a third point is chosen to define the radius of the central arc (usually taking the span, or a fraction of it, from the springing line as in figs. 16a and b). The method is quite simple and may be adjusted by trial-and-error to any ratio of span to height. Fig. 16c illustrates some solutions to "exams" by the apprentice masons in Augsburg. The problem was to define the oval centrings for irregular vaults. It is evident that they were using a well-established technique. In the drawings only the final result is shown, but this must have been preceded by some experiments. Werner Müller, a German scholar who has contributed decisively to our knowledge of Gothic *Baugeometrie* and the history of stereotomy [Müller 1990, 2002] called attention to this important document [Müller 1972].

In Spain, the oval arch also appears as an element of Gothic architecture in the fifteenth and sixteenth centuries, used in the portals of churches and civil buildings, as well as in the ribs of the surbased vaults which support the choir at the foot of Spanish parish churches. Palacios [2003] has made the first research in this most interested aspect of Spanish building geometry. Two photographs of this kind of vault in the entrance to the church of San Juan de los Reyes in Toledo are shown in fig. 17.

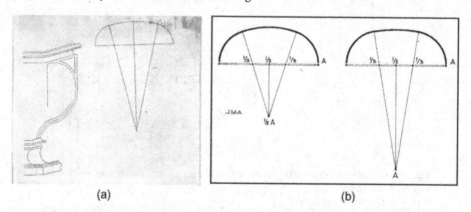

(a)　　　　　　　　　　　　　　　　　(b)

Fig. 16. Ovals in late Gothic German documents. a) Oval layout in the Wiener Sammlung [Koepf 1969]; b) Some types of Gothic ovals [Bucher 1972]

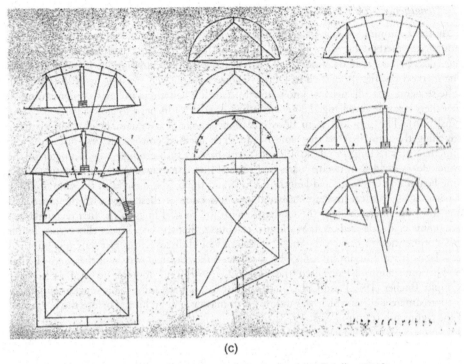

(c)

Fig. 16c. Exam of a mason apprentice in Augsburg [Müller 1972]

(a)   (b)

Fig. 17 . Gothic surbased vaults with oval ribs in the church of San Juan de los Reyes
in Toledo, designed by Juan Guas (end of the fifteenth century). a) Vaults at the
entrance (photo by G. López); b) Vault of the choir (photo by E. Rabasa)

The "lengthened arch"

At the beginning of the sixteenth century appears documentary evidence of another geometrical method for drawing oval arches or lines. It is first included in the *Codex Atlanticus* by Leonardo da Vinci (ca. 1510), as has been noted by Simona [2005] (fig. 18a). The method was first published in Dürer=s *Unterweisung der Messung* in 1525 (fig. 18b): "The stonemasons will need to know how to stretch a half circle in length, maintaining all the others measures unchanged, and this is because the vaults need to close adequately" (my translation). The problem is to draw a surbased arch of a certain height and span by stretching the semicircle of the same height. The method consists in first inscribing the semicircle in a rectangle of ratio 1:2; next, the base of the rectangle (the diameter of the semicircle) is divided in twelve parts and vertical lines are drawn. Another rectangle of the same height and of the desired length is drawn and the base is divided in the same number of parts; again, vertical lines are drawn. The intersection of these lines with the horizontal lines from the intersection of the semicircle with the vertical lines in the first rectangle will give points of the desired arch. We know, of course, that the curve is an ellipse, but Dürer does not mention it; in fact, the mathematicians would discover this fact only one hundred years later. It was Guldin in 1640 who discovered the elliptical nature of the curve [Peiffer 1995]. Some authors have used the term "pseudo-ellipse" to refer to oval arches (for example, Bucher [1972] and Calvo [2002]), but this is misleading, as it suggests the desire to approximate a curve which was unknown to masons and architects before the beginning of the eighteenth century [West 1978].

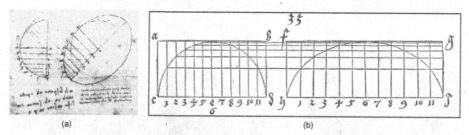

(a)   (b)

Fig. 18. Drawing of an oval by stretching a circle. a) Leonardo da Vinci, *Codex Atlanticus* fol. 318 b-r, ca. 1510 [Simona 2005]; b) Albrecht Dürer, *Unterweisung der Messung* [1525]

Is this construction Leonardo=s or Durër=s invention, or are they describing a method pertaining to the tradition of the stonemasons? In the late Gothic drawings which have survived there is no trace of such a geometrical construction; as we have seen, they solved the problem by using ovals made of circular arcs. However, it is interesting to note that the same construction appears in other architectural and stone-cutting treatises written in the last half of the sixteenth century in Spain, namely those of Hernán Ruiz [ca. 1545], Ginés Martínez de Aranda [1590] and Alonso de Vandelvira [1580]. As procedures and building methods change very slowly, it is probable that the method was one of the geometrical tools of medieval master masons.

The oldest of these treatises is the so-called *Libro de arquitectura* of Hernán Ruiz el Joven, written between 1545 and 1562 [Navascués 1975]. It is a manuscript containing a translation of Book I of Vitruvius, followed by a collection of drawings of different problems of geometry, stone-cutting, buttress design, classical orders and designs of temples. Hernán Ruiz includes several drawings showing a method equivalent to that of Dürer but based on the parallel projection on an inclined plane. The method is applied three times: to draw the caissons of a *capilla por arista*, groined vault (fig. 19a); to draw a *capilla vaída*, sail vault (fig. 19b); and to derive a surbased arch from a semicircular arch (fig. 19c). The use in three different cases and the absence of any instructions suggest that the method was well known among Spanish stonemasons in the sixteenth century.

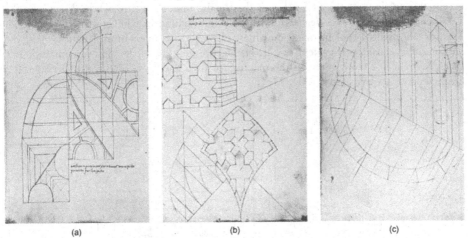

(a)   (b)   (c)

Fig. 19. The method of parallel projection in the *Libro de arquitectura* of Hernán Ruiz el Joven [ca. 1545]. a) Groined vault; b) Sail vault; c) arch construction. (Library of the E. T. S. de Arquitectura de Madrid)

Hernán Ruiz made extensive use of Serlio=s treatise of architecture and he also copied Serlio=s oval constructions (which will be discussed below). The method is not explicitly mentioned by Serlio, though he also uses parallel projection to obtain the intersection of the two barrels in his perspective drawing of a groined vault (fig. 20).

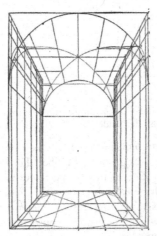

Fig. 20. Serlio=s perspective drawing of a groined vault using parallel projection [1996, Book II]

Martínez de Aranda, a Spanish architect who flourished at the end of the sixteenth century, wrote a treatise on stonecutting in about 1590 which has survived in manuscript copy [Martínez de Aranda 1986]. He gives two methods to obtain oval arches: one by contraction or extension of coordinates (equivalent to that of Dürer) and the other by parallel projection (equivalent to that of Hernán Ruiz) (fig. 21a and b). However, it is the second method which is used extensively throughout the treatise. Both methods are succinctly explained in the first pages of the treatise under the title "Difinitiones". In the stonecutting treatise of Vandelvira [1580] only the method of parallel projection is used, forming an essential tool in many drawings. However, he gives no explanation of it; the author either considers it evident or part of the common knowledge of a stonemason. The first occurrence of the method in the manuscript (fol. 13 r), where it is used to lay out a groined pendentive is shown in fig. 21c.

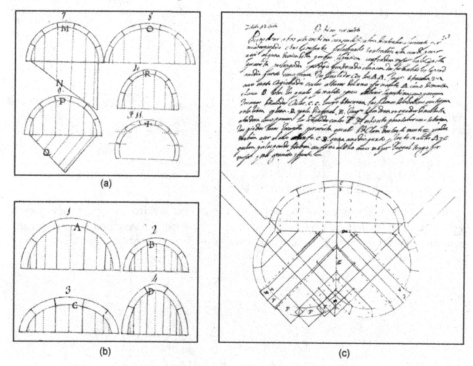

Fig. 21. The method of parallel projection for drawing oval arches: a and b) Martínez de Aranda [1590]; c) Alonso de Vandelvira [1580]

A long and detailed explanation of the two methods, a step-by-step exposition which can be followed by any apprentice, was given by Philibert De L=Orme in his book *Nouvelles inventions pour bien bastir* [1561]. A more succinct exposition is given in his book on architecture [1567]. De L=Orme calls the oval arch *Cherche rallongée*, lengthened arch, and again it seems that he is explaining a well known and established technique. De L'Orme gives two solutions which correspond approximately to the methods explained above (figs. 22a and b). He says explicitly that the curve cannot be draw with a compass and he shows the way to obtain the local curvature of the arch so that the templates for cutting the stones can be made.

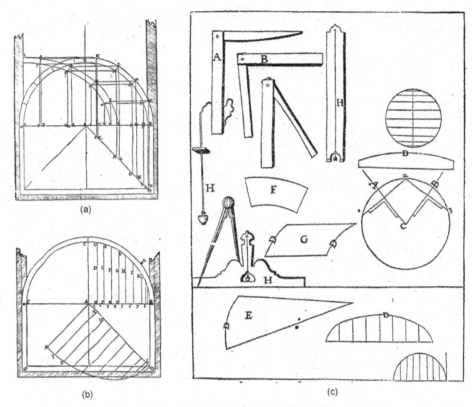

(a)

(b)

(c)

Fig. 22. Philibert De L=Orme's methods for constructing oval arches [1561, 1567]. a and b) two geometrical constructions; c) finding the centre of a circle when three points on the circumference are known

It is very simple: you take three successive points and draw two straight lines joining them. Then draw the perpendicular bisectors of each of these lines. The point of intersection of the two bisectors the centre of the circle which passes through the three points, as is shown in the sketch at the right of fig. 22c.

The same procedures were used by carpenters and were explained by Jousse in the first published handbook on carpentry printed in 1627 [1702] (fig. 23, next page). Of particular note is the drawing of median lines to obtain the radius of different parts of the arch (fig. 23c). It seems evident that a simplification to an oval of a few centres will be quite easy.

The oval in the Renaissance

We have seen that ovals were used in European late Gothic architecture. We have also seen that stonemasons developed a technique to lay out oval arches of any ratio of span to height by stretching or shortening circles or quarter-circles. The oval was not itself an invention of the Renaissance, but the oval form was used by some Renaissance architects to design a new concept of space for temples. Architectural historians have discussed the matter of the appearance of the oval in great detail; the fundamental contribution is still Lotz's monograph published in 1955. The literature is quite extensive; to name but a few fundamental contributions: Fasolo [1931], Zocca [1946], Müller [1967], Kitao [1974] and Nobile [1996]. On oval domes in Spain, see Gentil [1996] and Rodríguez de Ceballos [1983, 1990].

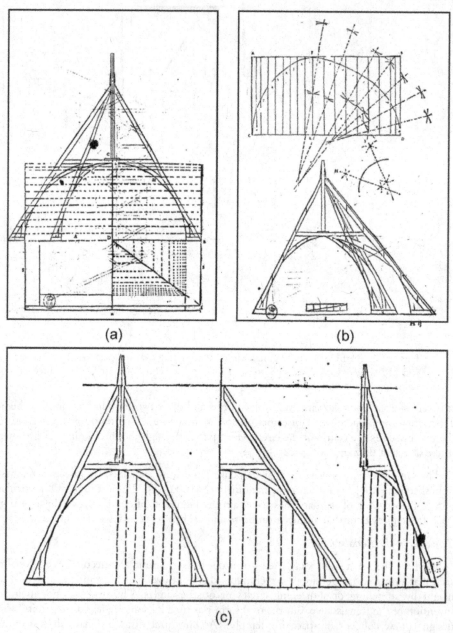

(a)

(b)

(c)

Fig. 23. The "lengthened arch" method in the manual of carpentry of Jousse [1627]

Apparently the idea of using the oval in different aspects of the arts was "in the air" at the beginning of the Cinquecento. Panofsky [1937] argued that Michelangelo's first project for the Tomb of Julius II already contained an interior oval space. According to Panofsky [1956], Correggio was the first painter to introduce an oval in a composition (*Madonna of St. Francis*, 1514, Dresden Gemäldegalerie), and Gian Maria Falconetto the first sculptor to employ one. In the book on perspective of Pélerin [1521] appear two ovals: the perspective drawing of an oval Renaissance arch and the oval frame of the illustration at the end of the treatise (fig. 24). It seems clear that the oval form exerted a new attraction to the artists at the beginning of the Cinquecento.

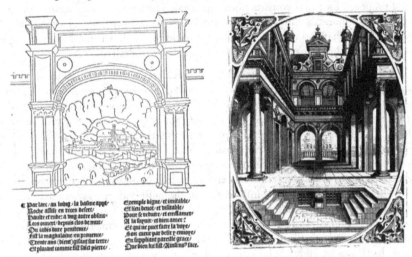

Fig. 24. Drawings with ovals in the treatise on perspective of Pélerin [1521]

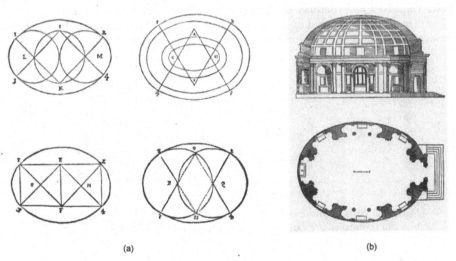

(a) (b)

Fig. 25. a) Serlio's models for ovals in his Book I of 1545; b) Serlio's design of an oval temple in his Book V of 1547

The main architects in promoting the oval as a new form of defining the architectural space were Baldassare Peruzzi, Sebastiano Serlio and Giacomo Vignola. It was Peruzzi who first thought in taking advantage from the peculiarities of an oval space in church design, a compromise between the central space of the Quattrocento and the more linear character of traditional churches. However, his death left the diffusion of his ideas in the hands of his disciple Serlio. Indeed, Serlio's treatise [1996], one of the most popular architectural treatises ever published, was responsible of the spread of the oval form in the late Renaissance and Baroque in Europe. In his Book I on geometry, published in 1545, he includes a discussion on ovals. He says explicitly that it is possible to draw many different ovals and proposed four oval constructions (fig. 25a). These were copied again and again in later architectural manuals and were used many times in actual designs. However, architects and masons knew that for any two axes it is possible to construct many (in fact, infinite) different ovals and they departed from Serlio's models when desired.

Serlio includes another construction for ovals. He first alludes to the practice of masons of describing the ovals with strings: *molti muratore hanno una certa sua prattica, che col filo fanno simili volte*. The method alluded by Serlio has been considered by many scholars to be the so-called "gardeners-method" for drawing the ellipse with a string fixed in the two foci. Kitao [1974] provides another, simpler interpretation (fig. 26a). Further on, Serlio proposes another method, which he considers more precise: *Nondimeno se l'architetto vorrà procedere teoricamente, portato dalla ragione, potrà tener questa via*. He then gives a different method of projection. Two concentric circles are drawn and equally-spaced radii drawn to partition them. From the intersection of the radii with the larger circle are drawn vertical lines; from the intersection of the radii with the smaller circle are drawn horizontal lines. The intersections of these verticals and horizontals are points on the oval (fig. 26b). Of course, the curve is an ellipse, but Serlio did not name it so; more importantly, he did not know that it was an ellipse. This construction was also copied in subsequent treatises of the sixteenth and seventeenth centuries; after that, it gradually disappears from the manuals, although it continues to appear in the handbooks of practical geometry. (Hernán Ruiz, Vandelvira and Martínez de Aranda include it. However, in the usual stonecutting constructions they never used it; it only appears at the end of the manuscript by Vandelvira, in what looks as a theoretical exercise).

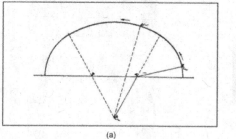
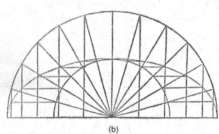

(a)   (b)

Fig. 26. a) Physical construction of an oval with a string and two pegs [Kitao 1974];
b) Serlio's construction of an oval *portato dalla ragione* [Serlio 1545]

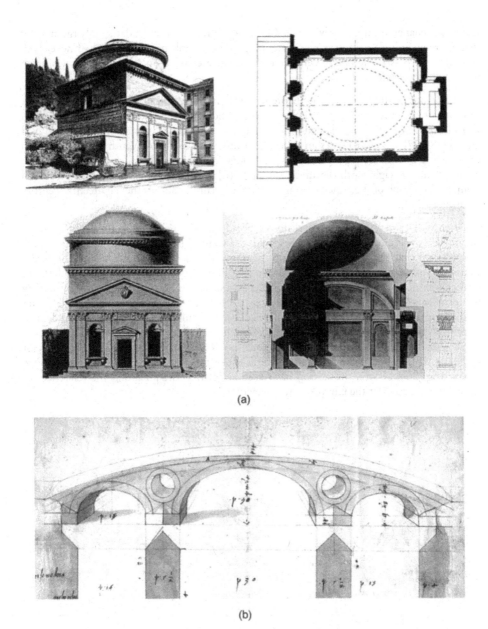

(a)

(b)

Fig. 27. a) The first church built with an oval dome. S. Andrea in Via Flaminia by
Vignola (1550-1554); b) First design of a bridge with oval arches by Vignola in 1547

When Serlio included the oval constructions in his Book I he was not thinking only of a
geometrical problem; he had in mind the program of his master Peruzzi, the creation of a
new type of temple. In his Book V on architecture, published two years later [1547],
among a number of projects for temples, there appeared a design for an oval temple (fig.
25b), which exerted an enormous influence in the late Renaissance and Baroque. This is an

enormous change from the previous uses of the oval: it became a fundamental argument of design and acquired an importance in defining architectural space which it had never had. Indeed, it surpassed the circle, which had been since classical antiquity the expression of geometrical perfection.

Peruzzi and Serlio prepared the theoretical way. It was Vignola who first put their ideas into practice building the first oval church, Sant'Andrea in Via Flaminia (1550-1554) (fig. 27a). Earlier, in 1547, Vignola had produced the first design of a bridge with surbased oval arches (fig. 27b). Vignola's design exerted no influence in the sixteenth and seventeenth centuries, but the use of oval arches in bridge design had become common by the beginning of the eighteenth century [Gautier 1716]. Polycentric oval arches were a central element in many of Perronet's bridges, in the second half of the eighteenth century [Perronet 1788].

After Serlio and Vignola, the oval dome spread quickly, not only in Italy, but in Spain [Gentil 1996], France [Châtelet-Lange 1978] and Central Europe [Fasolo 1931] as well. It is significant that Vandelvira's manuscript of stone cutting, ca. 1575, includes six different solutions of oval domes. There is no space in this short paper to enter into the subtleties of oval church design. The best architects of the Baroque exercised their ingenuity by solving the problems created by a non-central space. For example, Smyth-Pinney [1989] has studied in detail the design process followed by Bernini for the plan of S. Andrea al Quirinale and the subtle position of the axes of the perimeter chapels. Bernini's oval does not correspond to any of Serlio's models, and shows a delicate adjustment in the interior space (fig. 28). However, we may call Bernini's use of the oval "classical". As in Serlio's oval temple (fig. 25b, above), the section has the same form as the plan, and the references to the central churches of the Cinquecento are evident.

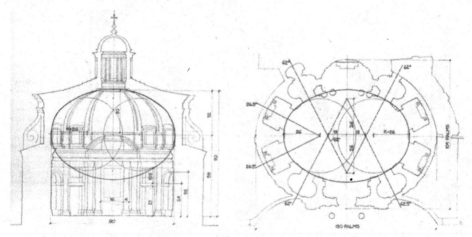

Fig. 28. The geometry of San Andrea al Quirinale [Smyth-Pinney 1989]

The project of Francesco Borromini for San Carlo alle Quattro Fontane (1663) presents a more complicated geometry, as the oval which generates the plan changes at the base of the dome (fig. 29). This last oval deviates very much from the usual form of ovals so far. No doubt, Borromini chose this form to provide "tension" in the space. Neither of the ovals corresponds with Serlio's models.

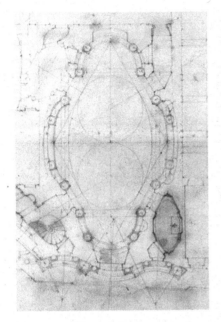 

Fig. 29. San Carlo alle Quattro Fontane by Borromini. a) Design for the plan using a generating oval; b) Photograph of the dome showing the oval of impost at the base of the dome. Note the difference with the oval plan [Bellini 2004]

Geometry of oval domes

One of the main problems in the study of the geometry of oval arches and domes is the modern prejudice that an "oval" is an "approximation of the ellipse". In fact, as we have seen, as a geometrical figure the oval is much older than the ellipse, and there is a tradition of its use in building practice which can be traced back to the first arches in Mesopotamia and Egypt. As a mathematical concept, in geometrical terms, the ellipse is an intersection of a plane with a cone, or the locus of the points whose distance to two fixed points (foci) give a constant number. It is also an affine transformation of a circle. And we may define the curve mathematically in many other ways.

The study of the conics was a matter of "higher mathematics" up to the seventeenth century. The Greeks did not know a simple way to draw the ellipse. The string method was discovered by Anthemius of Tralles, a Byzantine architect and geometer of the sixth century. In the tenth century the Arab Alsigzï reported it, and in the late sixteenth and early seventeenth centuries it was known by Guidobaldo del Monte, Simon Stevin and Kepler [West 1978]. As a way to lay out arches, it was published in the treatise of fortification of Bachot [1598] (fig. 30a), and it began to appear in the building manuals at the beginning of the seventeenth century. In Spain, Fray Lorenzo de San Nicolás [1639], who wrote the most influential building manual in the Spanish language, after explaining the usual layout with circular arcs, cites the method and remarks that it is easy to lay out brick arches (fig. 30b and c). However, in the chapter on ovals he explains in detail the usual circular arc constructions, adding some new constructions to those of Serlio.

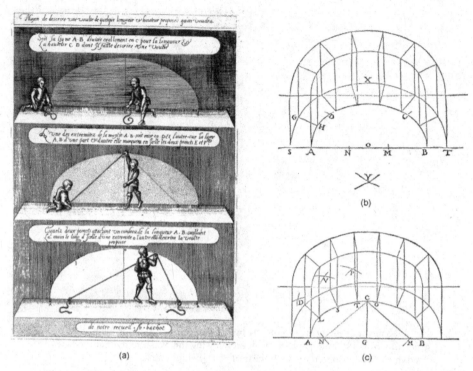

Fig. 30. a) Use of the string method to draw surbased arches in Bachot [1598]; b and c) Layout of surbased arches by employing circular arcs, and by the string method [San Nicolás 1639]

In the eighteenth century contempt for the oval on the part of learned engineers and architects began to grow. Frézier [1737] sharply criticized the use of ovals, and the title of Chapter IV of his Book II is illustrative: *De l'imitation des Courbes Régulieres par des compositions d'Arcs de Cercle*. He adds, *je ne conseille à personne d'avoir recours à cet artifice de l'ignorance*. However, he included a description of how to lay out different ovals. The treatise of geometry by Camus [1750] contains the most complete exposition of the different methods to lay out ovals for certain fixed conditions.

Thus it appears that in the mid-eighteenth century began a divorce between what is theoretically good and what was done in the practice. That a method is known does not necessarily imply that it must be used. The practical advantages of using circular arcs must have been taken into account. No doubt, Bernini and Borromini knew the methods of drawing ellipses, but they used ovals.

As was mentioned in the case of the Roman amphitheatres, some scholars believe that ascertaining whether the architect used an oval or an ellipse is a matter of mensuration. In at least one case, Rosin [2005] has demonstrated the practical impossibility of distinguishing an oval from an ellipse. Bendetti [1994] and Migliari [1995] arrived to the same conclusion after studying the geometry of the profile of the dome designed by Antonio de Sangallo for St. Peter's. Gentil [1996] claimed that the photogrammetrical survey of the oval dome of the Sala Capitular in the Cathedral of Seville demonstrated the use of ellipses, and other scholars agree with him [Rabasa 2000; Palacios 2003a]. The Sala has a ratio of length to width of 4:3, and we have seen in fig. 5a a geometrical construction for an oval of this proportion, the "Egyptian" oval. In fact, Vignola knew and used this constructions, as Gentil points out. For this geometry the difference between the ordinates of the oval and the corresponding ellipse amounts to less than 0.7% of the major span. It is not possible to reach this precision in building practice, and besides, the unavoidable movements suffered by masonry vaults and domes after decentring are also at least on the same order. We must arrive, then, at the same conclusion of the scholars cited above: determining the geometry of an actual oval of usual proportions is not a problem of mensuration. When Vigonla's oval is superimposed onto the reproduction of the photogrammetry in Gentil's article, the agreement is quite good (but of course, it is also with the ellipse) (fig. 31).

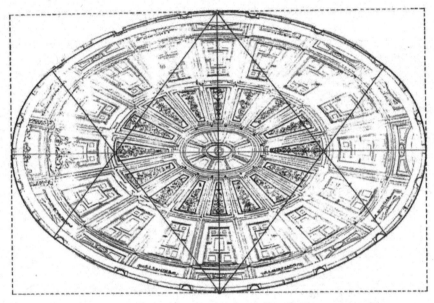

Fig. 31. The Sala Capitular of the Cathedral of Seville. Comparison of the photogrammetrical survey [Gentil 1996] with Vignola's 4:3 oval

To move forward towards solving the problem it is necessary to take into account the constructive tradition and the problems posed in practical building by the use of a curve such as the ellipse, which has a varying radius of curvature. In the case of a stone arch or dome, this will imply the use of different templates for each stone. We may also turn to documentary evidence. In the cases of S. Andrea and S. Carlo alle Quattro Fontane, the extant drawings show the employment of ovals. The same occurs in the case of the dome of Vicoforte [Aoki, et al. 2003]. In fact, this author does not know any plan of an oval dome of significant size (say a span of more than 10 m) that shows the use of the ellipse, although this does not mean that the ellipse may not have been used in exceptional cases. As an example, the Spanish architect Plo y Camin claimed in his treatise of architecture [1757] that he has laid out a brick oval dome with an ellipse compass formed by two wooden planks (no. 27 in fig. 32). This device can function only for domes of small size, as the bending of the planks will limit their size.

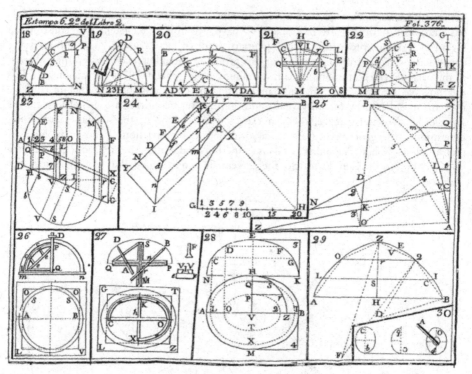

Fig. 32. Construction of oval domes in the treatise of Plo y Camín [1767]

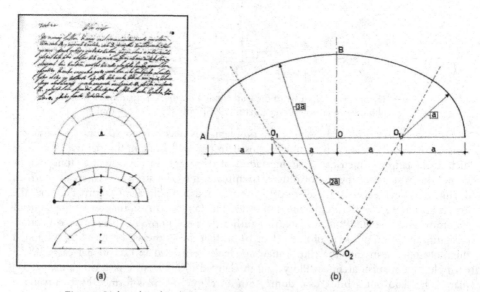

(a)   (b)

Fig. 33. Surbased oval arch (*arco carpanel*) of Alonso de Vandelvira (ca. 1580)

The oldest description of the geometry of oval domes is contained in the manuscript on stone cutting of Alonso de Vandelvira[1580]. There he explains six different types of *capillas ovales*, oval domes. All of them have the same plan – a *carpanel*, oval, of four centres – which Vandelvira uses througout his treatise (fig. 33). However, Vandelvira notes that other ovals can be chosen. The *capillas ovales tercera y cuarta*, third and fourth domes, provide a good opportunity for studying the geometry. Like the dome of the Sala Capitular (fig. 31), they are ribbed oval domes and the problem is to define the geometry of the "meridian" and "parallel" ribs.

Before discussing Vandelvira's construction some general comments on the building and geometry of oval domes should be made. The usual form of a dome has axial symmetry. In modern terms it may likened to a solid of revolution. For a master mason or architect, simply put, an arch defines the geometry of the dome. The advantages from the point of view of construction need not be explained. Besides, the dome may be erected without scaffolding, by closing successive rings of masonry. What is needed, then, are not centrings but "guides", a light scaffold which defines the geometry in space. Sometimes these arches were completed first and then the masonry shell between them was constructed. The advantage here is that the work can proceed faster, as the lime mortar of the arches will dry more quickly when four faces are exposed to the air as opposed to a single thick shell. Also, the placing of iron rings served to maintain the geometry while the mortar was setting. It was a complex process and there were several methods. It is necessary to keep in mind the remark of Rodrigo Gil de Hontañón when he described the closing of a cross vault: "... these things may be difficult to understand if one lacks experience and practice, or if one is not a stone mason, or has never been present at the closing of a rib vault" [García 1681]. The statement can be applied verbatim to the building of a masonry dome and reminds us of our basic ignorance about even the most simple operations of masonry construction.

The geometry of an oval dome is much more complex than that of the usual dome with a central vertical axis of symmetry. A modern architect or engineer will think of the usual ways of generating surface: rotation, translation, affine transformation, etc. The architect would have to think first in a general way of parameters which define the overall form of the intrados of the dome: the relationship between the two axes of the oval plan, the relationship between the height and the span, and the profile of the dome. All these parameters must have a certain relation with one another. Then comes the problem of how to build the dome, that is, how the surface of the intrados is going to be defined in the space, perhaps by a series of centrings. If the dome is to be made of stone, the geometrical definition must take into account the cutting of the stones. All the processes of building are subtended by a principle of economy.

A simple way of defining the geometry is to fix the profiles of the two sections following the major and minor axis of the oval plan. In Serlio=s design of the oval temple (fig. 25, above), the curve which forms the section of the dome is the same oval as that of the plan. Because the height is the same as half the minor axis, the transverse cross section will be a circle, the simplest option. Any other transverse section is also a semicircle and by placing semicircular transverse centring the dome may easily built by successive rings, until it is closed. For a modern architect or engineer this is, of course, a surface of revolution around the major axis, but it is quite improbable that Serlio would have thought of it in such a way. Many Renaissance and Baroque oval domes have this kind of geometry. See, for example, San Andrea del Quirinale (fig. 28).

The same approach may applied considering that the transverse section has the form of the semi-oval and that the longitudinal section is a semicircle. This was the approach taken in Sant'Andrea in Via Flaminia (fig. 27a) and in the dome of the Cesarean Library in Vienna (fig. 34).

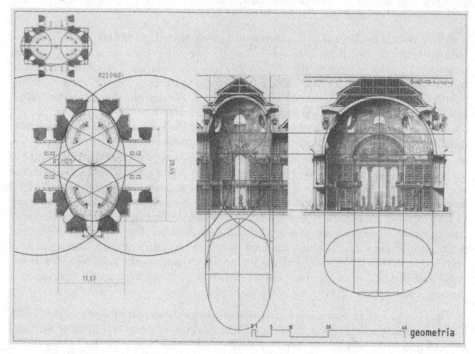

Fig. 34. Geometry of the oval dome of the Cesarean Library in Vienna (drawing by
C. Pérez de los Ríos, after G. López Manzanares [2005])

Again, the logical placement of the successive centrings for the building of the dome will be the transverse sections, but what is the form of the shorter transverse sections? Now, it is not as simple as in the previous case, in which a semicircle can be supposed. It will be simple to think that these sections have the same form of the transverse minor section, that is, that they are similar ovals of diminishing size. To fulfil this condition the vertical ordinates of the circle and the oval must have a constant relationship, i.e., the oval is an ellipse. This is the mathematical point of view. Thus it follows that the oval plan must be elliptical to permit economical building. The point of view of the builder is that the oval fulfils the condition within the tolerances of building, something of the order of 1-3% of the span. Indeed, the deviation of the ordinates of the ovals used in building is less than this. Therefore the false supposition that transverse sections are all similar becomes "practically true". Building manuals are full of false "practical truths", to approximate values (for example, the square root of 2 is approximated to 7/5 with an error of 1% in the late Gothic German architectural manuals, and the practical value of π is 22/7, with less than 0.1% error [Huerta 2004]. The standards of surveying and levelling could not guarantee more precision and therefore the use of approximate numerical values and geometrical constructions is not only fully justified, but shows how clever the old master builders were.

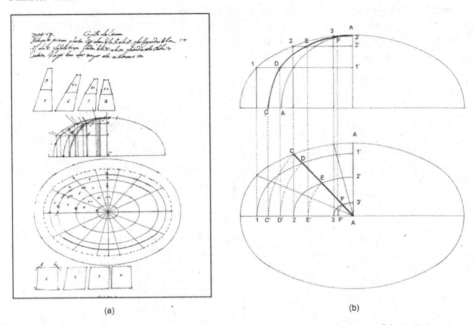

Fig. 35. a) Third oval dome (*capilla oval tercera*) of Vandelvira (Library of the E.T.S. de Arquitectura de Madrid); b) Method for drawing a "meridian rib"

Now, in order to discuss Vandelvira's solution for oval domes, we should keep in mind that he was not looking for the exact solution of the problem but rather for a practical solution that was easy to construct and draw. The process is clear with reference of the third oval dome (fig. 35a). He is considering an oval dome with the same longitudinal section as half the oval plan. Further, the cross section is a semicircle A-A. The longitudinal section is divided in seven parts by points 1, 2, 3, etc. (we work with only half section). Then horizontal lines are drawn and points 1', 2', 3' are obtained on the semicircular cross section. Also, verticals from points 1, 2, 3, are projected onto the plan. Then, in the plan, we know that an horizontal section of the dome surface will give a line which must pass through points 1-1', 2-2', etc. What is this curve? If the surface is generated by the revolution of the oval plan around the long axis, we may draw this line point by point. Or, if the plan is supposed be an ellipse, then, it will be an ellipse. But the first procedure would be too long and Vandelvira did not know the second property. He then assumes that every horizontal section has a form similar to the oval of the plan. Of course, this is not true, as the proportion between the semi-axes A1-A1', A2-A2', ..., need not coincide with the proportion of the semi-axes of the oval plan. But the error is small in practice, and even the drawing can be "forced" by slightly opening or closing the compass, as the present author has done to make the drawing (a practice that any student of pre-CAD technical drawing of times knows very well; one of the problems of drawing by computer is that it is very difficult to "trick" a solution). Is Vandelvira ignorant? Frézier would have said this. But in fact, he is very clever: he is using a very simple graphical construction to solve a problem that, expressed in rigorous geometrical terms, would have been impossible. He is not a mathematician; he is a builder. In fact the approximation functions because of the small difference between ovals and ellipses alluded to. The construction will be exact for an ellipsoid of revolution, but he did not need to know this. Vandelvira uses the same approach in the solution of fourth oval dome (fig. 36) and, most probably, the Sala Capitular in Sevilla would have been laid out following the same method.

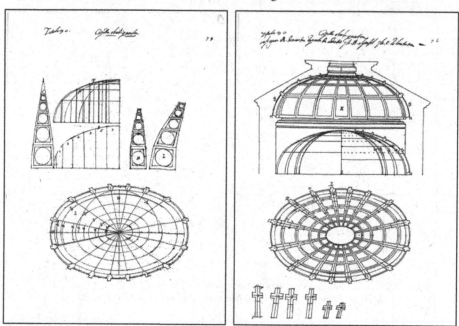

Fig. 36. Fourth oval dome (*capilla oval cuarta*) of Vandelvira (Library of the E.T.S. de Arquitectura de Madrid)

Mechanics of oval arches and domes

An "arch" is the natural way to span a void with a material in compression only, i.e., piling stones of a certain shape, so that the resulting geometry is stable: the stones, wanting to fall due to the force of gravity, remain in place due to their mutual interaction.

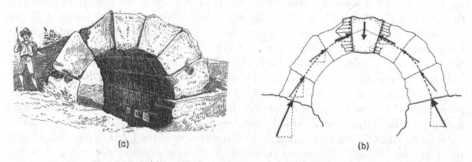

(a)   (b)

Fig. 37. Equilibrium of a voussoir masonry arch [Huerta 2004]

In fig. 37 we can see a massive Etruscan voussoir arch. Heavy stones have been cut in the form of wedges; then they have been placed on a formwork (a centring) beginning from the extremes. When the last stone in the middle (the keystone) was placed the centring was removed and the arch stood. (The best exposition of the mechanics of masonry structures is [Heyman 1995]).

Let us consider the free-body equilibrium of the keystone. In each joint (which we imagine more or less planar) there will exist a certain stress distribution. The stress resultant must be a compressive force, a "thrust"; the point of application is the "centre of thrust" and it must be contained within the plane of the joint. The two thrusts in the joints maintain the keystone in equilibrium. The same occurs with the other stones until we arrive at the springing of the arch. There the abutment must supply/resist a certain thrust. This is the "thrust of the arch", and the abutment must have adequate dimensions to resist it.

The masonry arch always pushes outwards; "the arch never sleeps", says an old proverb attributed to the Arabs. The locus of the centre of thrust forms a line, the "line of thrust". The form of this line depends, therefore, on the geometry of the arch. Of course, as masonry must work in compression, the line of thrust must be contained within the masonry arch.

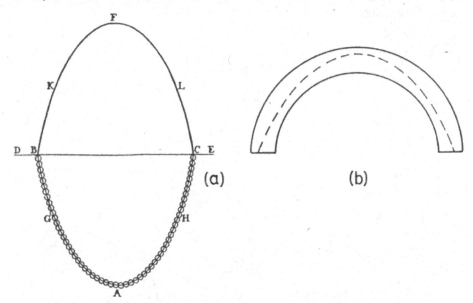

Fig. 38. Possible line of thrust of a semicircular arch [Heyman, 1995]

We may imagine one voussoir acting against the two voussoirs adjacent to it only through the centres of thrust. If we now invert the arch, what was a force of compression will be a force of tension: the voussoirs are hanging like a chain. The statics of arches and hanging cables is the same. (This was Hooke's brilliant idea, ca. 1670.) If all the voussoirs of the arch are the same size the line of thrust will have very nearly the form of an inverted catenary. But, it is not necessary that the arch have the form of catenary: it suffices that the catenary can be contained within the masonry. Therefore, in fig. 38 the arch is in equilibrium with an internal stress distribution represented by the inverted catenary. The drawing has no scale and it is evident that the safety is a matter of geometry, independent of size [Heyman 1995; Huerta 2006].

A dome can be imagined as composed by a series of arches obtained slicing the dome by meridian planes. Every two "orange slices" form an arch; if it is possible to draw a line of thrust within this arch, then we have found a possible equilibrium state in compression and the dome is safe, that is, it will not collapse (fig. 39a) [Heyman 1995]. The dome may have an oculus because a compression ring forms: the dome build a "keystone" when a ring is closed, and therefore masonry domes can be built without centring The safety condition is, then, a geometrical condition. Domes of similar forms and different sizes have the dame degree of safety: the drawing of fig. 39a has no scale.

The same "slicing technique" can be applied to oval domes [Huerta 2004]. Now the elementary arches are different, but equal the two opposed which can build a safe arch. If it is possible to find a line of thrust within each pair of elementary arches, the dome will be safe (fig. 39b). The only difference is that, since the arches are different, the compression ring has an oval form. As the oval form of the compressions need not coincide with the oval form of the oculus, the oculus must have sufficient thickness to accommodate the compression ring within the masonry. Again, the safety depends on geometry and not on the size of the dome.

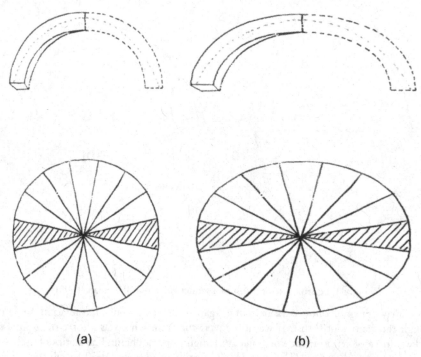

(a)   (b)

Fig. 39. Use of the "slicing technique" to study the safety of masonry domes. a) Domes of revolution; b) Oval domes

Scaling up and down is a particular case of an affine transformation. Rankine [1858] discovered that the stability of masonry structures subject only to dead load remained unaltered after an affine transformation: the line of thrust of the transformed arch is the affine transformation of the original line. The relative distances to the limits of masonry does not change, and, therefore, the safety is the same (fig. 40). (For a detailed discussion see [Huerta 2004]).

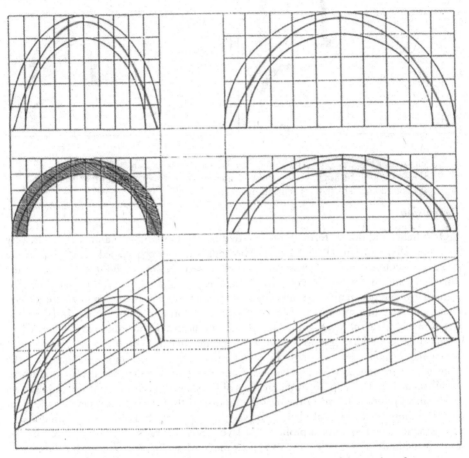

Fig. 40. Affine transformations of a masonry arch and its line of thrust: the safety remains the same [Huerta 2004]

The same occurs with domes. If we transform a stable dome by an affinity, the transformed dome has the same degree of safety and, again, the surface of thrust, representing the internal equilibrium, is the transformation of the original surface of thrust. Fig. 41a represents the statical analysis of the model of a simple dome proposed by Fontana [1694]. Fig. 41b represents another dome, obtained by contracting the height and width of the original dome. The line of thrust represents the equilibrium of the slice arch having the major axis. The stability remains unaltered. It may be that the confidence of the architects in the stability of oval domes originates in the intuition of this principle.

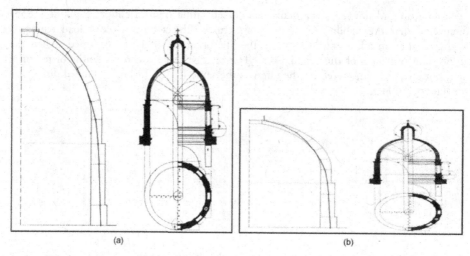

(a)

(b)

Fig. 41. a) Design of a simple dome by Fontana [1694] and analysis of its stability [Huerta 2004]; b) Safe oval dome obtained contracting the height and broad of the previous dome. The stability remains unaltered

Conclusion

Oval arches and domes form part of the tradition of building in masonry from the very beginning of the invention of the arch. The approximately egg-shaped forms were made regular through the use of practical geometry, perhaps about 2000 B.C., when the dimensions and the importance of the construction required it. The oval forms, generated by the tangency of circular arcs remained as an essential part of the architecture of Asia Minor and were incorporated in Islamic architecture, forming polycentric pointed arches. The Romans used the oval to design their amphitheatres. The oval arch reappeared in Europe in the Middle Ages as a device for solving practical problems: a dome over a rectangular bay, a surbased arch. In late Gothic, the master builders broadened the geometrical possibilities by employing oval ribs. They probably deduced the method of the "lengthened arch" that appears in the treatises of the sixteenth century. In the Cinquecento, Renaissance architects found in the oval the motif and the solution for a new architectural type: the church with an oval plan. This type was much appreciated during Baroque and late Baroque. The geometrical problems of laying out an oval surface were solved by the simplest methods, sufficiently close in approximation for practice. Though the ellipse was incorporated into building treatises in the seventeenth century, it was rarely used, and was probably never used at all in stone domes.

The mechanics of an oval dome is analogous to a dome of revolution, and the same techniques of analysis can be used. The dome will be stable if it is possible to find a thrust surface, in equilibrium with the loads, within the masonry. This is a matter of geometry. It turns out, that a "lengthened arch" or a "lengthened dome" has the same stability as the original, undistorted, arch or dome. Rankine demonstrated this property in the mid-nineteenth century, but it is obvious that the architects knew this property intuitively. This intuition was probably the basis for the confidence of architects who designed and constructed large oval domes from the beginning.

The study of the influence of mathematics in building cannot be made without considering the essential objective of a builder: to build. The complication of the geometry was always tempered by the common sense of the builder and the imperative necessity to erect a safe building in a certain time. It is true that architects enjoyed the play of geometry, but play presupposes some rules, and they found their liberty within the rules of construction.

References

AOKI, T., M.A. CHIORINO, and R. ROCATTI. 2003. Structural characteristics of the elliptical masonry dome of the Sanctuary of Vicoforte. Pp. 203-212 in *Proceedings of the First International Congress on Construction History*, S. Huerta, ed. Madrid: Instituto Juan de Herrera.

ARNOLD, Dieter. 1991. *Building in Egypt. Pharaonic stone masonry*. Oxford: Oxford University Press.

BACHOT, Ambroise. 1598. *Le Gouvernail d'Ambroise Bachot capitaine ingenieur du Roy, Lequel conduiraie curieux de Geometrie en perspective dedans l'architecture des fortifications, machines de guerre et plusieurs autres particularites et contenues.* Melun : Chez l'Auteur.

BAIMATOVA, Nasiba. 2002. Die Kunst des Wölbens in Mittelasien. Lehmziegelgewölbe (4.-3. Jh. v. Chr. - 8. Jh. n. Chr.). Dissertation: Institut für Vorderasiatische Altertumskunde, Freie Universität Berlin.

BELLINI, Federico. 2004. *Le cupole di Borromini: la "scientia" costruttiva in età barocca.* Milan: Electa.

BENEDETTI, Sandro. 1994. Oltre l'antico e il gotico. Il profilo della cupola vaticana di Antonio da Sangallo il Giovane. *Palladio* **14**: 157-166.

BESENVAL, Roland. 1984. *Technologie de la voûte dans l'Orient Ancien.* 2 vols. Paris: Editions Recherche sur les Civilisations.

BOYD, Thomas D. 1978. The arch and the vault in Greek architecture. *American Journal of Archaeology* **82**: 83-100.

BUCHER, François. 1968. Design in Gothic Architecture. A Preliminary Assessment. *Journal of the Society of Architectural Historians* **27**: 49-71.

BUCHER, François. 1972. Medieval Architectural Design Methods, 800-1560. *Gesta* **11**, 2: 37-51.

CALVO LÓPEZ, José. 2002. La semielipse peraltada. Arquitectura, geometría y mecánica en las últimas décadas del siglo XVI. Pp. 417-435 in *Actas del Simposium El Monasterio del Escorial y la Arquitectura* (San Lorenzo del Escorial, 8 al 11 de noviembre de 2002). San Lorenzo del Escorial.

CAMUS, M. 1750. *Elémens de géométrie théorique et pratique (Cours de mathématique, Seconde Partie).* Paris: Durand.

CARAZO, Eduardo and Juan Miguel OTXOTORENA. 1994. *Arquitecturas centralizadas. El espacio sacro de planta central: diez ejemplos en Castilla y León.* Valladolid: Servicio de Publicaciones de la Universidad de Valladolid.

CEJKA, Jan. 1978. Tonnengewölbe und Bogen islamischer Architektur. Wölbungstechnik und Form. Dissertation: München. Techn. Univ. Fachbereich Architektur.

CHAPPUIS, R. 1976. Utilisation du tracé ovale dans l'architecture des églises romanes. *Bulletin Monumental* **134**: 7-36.

CHOISY, Auguste. 1883. *L'art de bâtir chez les Byzantines.* Paris: Librairie de la Societé Anonyme de Publications Périodiques.

———.1904. *L'art de bâtir chez les égyptiens.* Paris: E. Rouveyre.

———. 1904. Note sur deux épures égyptiennes conservés à Edfou. *Journal of the Royal Institute of British Architects* **11**: 503-505.

DARESSY, Georges. 1907. Un tracé egyptienne d'une voûte elliptique. *Annales du Service des Antiquités de l'Egupte* **8**: 234-241.

DE L'ORME, Philibert. 1561. *Nouvelles inventions pour bien bastir et a petits fraiz.* Paris: Morel.

———. 1567. *Le premier tome de l'Architecture.* Paris: Morel.

DORNISCH, Klaus. 1992. *Die griechischen Bogentore. Zur Entstehung und Verbreitung des griechischen Keilsteingewölbes.* Frankfurt am Main: Peter Lang.

DÜRER, Albrecht. 1525. *Unterweisung der Messung.* Nürnberg. (facsimile edition, Nördligen: A. Uhl, 1983).

EL-NAGGAR, Salah. 1999. *Les voûtes dans l'architecture de l'Égypte ancienne.* Le Caire: Institut Français d'Archéologie Orientale.

FASOLO, Vincenzo. 1931. Sistemi ellittici nell'architettura. *Architettura e Arti Decorative* 7: 309-324.

FATHY, Hassan. 1976. *Architecture for the Poor. An Experiment in Rural Egypt.* Chicago/London: University of Chicago Press.

FONTANA, Carlo. 1694. *Il tempio Vaticano e sua origine.* Rome : Nella Stamparia di Gio : Francesco Buagni.

FREZIER, Amédée-François. 1737-39. *La théorie et la pratique de la coupe de pierres et des bois pour la construction des voûtes et autres parties des bâtiments civils et militaires, ou traité de stéréotomie à l'usage de l'architecture.* Strasbourg/Paris: Charles-Antoine Jombert.

GAUTIER, Hubert. 1716. *Traité des Ponts.* Paris: Cailleau.

GENTIL BALDRICH, José María. 1996. La traza oval y la sala capitular de la catedral de Sevilla. Una aproximación geométrica. Pp. 77-147 in *Quatro edificios sevillanos.* J. A. Ruiz de la Rosa et al., eds. Seville: Colegio Oficial de Arquitectos de Andalucía, Demarc. Occidental.

GÓMEZ-MORENO, Manuel. 1919. *Iglesias Mozárabes. Arte español de los siglos IX al XI.* Madrid: Centro de Estudios Históricos.

GÖTZ, Wolfgang. 1968. *Zentralbau und Zentralbautendenz in der gotischen Architektur.* Berlin: Gebr. Mann Verlag.

GOYON, J.-C., J.-C. GOLVIN, C. SIMON-BOIDOT and G. MARTINET. 2004. *La construction Pharaonique, du Moyen Empire à l'époque gréco-romaine.* Paris: Picard.

HEATH, Thomas. 1981. *A History of Greek Mathematics.* 1921. Reprint, New York: Dover Publications.

HEISEL, Joachim P. 1993. *Antike Bauzeichnungen.* Darmstadt: Wissenschaftliche Buchgesellschaft.

HEYMAN, Jacques. 1995. *The Stone Skeleton. Structural Engineering of Masonry Architecture.* Cambridge: Cambridge University Press.

HUERTA, Santiago. 2004. *Arcos, bóvedas y cúpulas. Geometría y equilibrio en el cálculo tradicional de estructuras de fábrica.* Madrid: Instituto Juan de Herrara.

———. 2006. Galileo was wrong! The geometrical design of masonry arches. *Nexus Network Journal* 8: 25-52.

JOUSSE, Mathurin. 1702. *L'art de Charpenterie . . . corrigé et augmenté . . . par M. de La Hire.* 1627. Paris: Thomas Moette.

KITAO, Timothy K. 1974. *Circle and Oval in the Square of Saint Peter's. Bernini's Art of Planning.* New York. New York University Press.

KOEPF, Hans. 1969. *Die gotischen Planrisse der Wiener Sammlungen.* Vienna: Hermann Böhlaus Nachf.

KRAUTHEIMER, Richard. 1984. *Arquitectura paleocristiana e bizantina.* Madrid: Cátedra.

LÓPEZ MANZANARES, Gema. 2005. La contribución de R. G. Boscovich al desarrollo de la teoría de cúpulas: el informe sobre la Biblioteca Cesarea de Viena. Pp. 655-665 in *Actas del Cuarto Congreso Nacional de Historia de la Construcción* (Cádiz, 27-29 January 2005). S. Huerta, ed. Madrid: Instituto Juan de Herrera.

LOTZ, Wofgang. 1955. Die ovalen Kirchenräume des Cinquecento. *Römisches Jahrbuch für Kunstgeschichte* 7: 7-99.

LUCIANI, Roberto. 1993. *El Coliseo.* Madrid: Anaya.

MARTÍNEZ DE ARANDA, Ginés. [ca. 1590] *Cerramientos y trazas de montea.* Ms. Biblioteca de Ingenieros del Ejército de Madrid.

———. 1986. *Cerramientos y trazas de montea.* Madrid: Servicio Histórico Militar, CEHOPU.

MIGLIARI, Riccardo. 1995. Elissi e ovali: epilogo di un conflitto. *Palladio* 8, 16: 93-102.

MÜLLER, Johann Heinrich. 1967. *Das regulierte Oval. Zu den Ovalkonstruktionen im Primo Libro di Architettura des Sebastiano Serlio, ihrem architekturtheoretischen Hintergrund und ihrer Bedeutung für die Ovalbau-Praxis von ca. 1520 bis 1640*. Bremen.

MÜLLER, Werner. 1971. Der elliptische Korbbogen in der Architekturtheorie von Dürer bis Frézier. *Technikgeschichte* **38**: 93-106.

———. 1972. Die Lehrbogenkonstruktion in den Proberissen der Augsburger Mauermeister aus den Jahren 1553-1723 und die gleichzeitige französische Theorie. *Architectura* **2**: 17-33.

———. 1990. *Grundlagen gotischer Bautechnik*. München: Deutscher Kunstverlag.

———. 2002. *Von Guarino Guarini bis Balthasar Neumann. Zum Verständnis barocker Raumkunst*. Petersberg: Michael Imhof Verlag.

NAVASCUÉS PALACIO, Pedro. 1974. *El libro de arquitectura de Hernán Ruiz el Joven. Estudio y edición crítica*. Madrid: Escuela Técnica Superior de Arquitectura.

NOBILE. Marco Rosario. 1996. Chiese a pianta ovale tra Controriforma e Barocco: Il ruolo degli ordini religiosi. *Palladio* **9**, 17: 41-50.

PALACIOS GONZALO, José Carlos. 2003. *Trazas y cortes de cantería en el Renacimiento español*. Madrid: Munilla-Lería.

———. 2003. Spanish ribbed vaults in the 15th and 16th centuries. Pp. 1547-58 in *Proceedings of the First International Congress on Construction History*, S. Huerta, ed. Madrid: Instituto Juan de Herrera.

PANOFSKY, Erwin. 1937. The First Two Projects of Michelangelo's Tomb of Julius II. *Art Bulletin* **19**: 561-579.

———. 1956. Galileo as a Critic of the Arts: Aesthetic Attitude and Scientific Thought. *Isis* **47**: 3-15.

PELERIN, Jean. 1521. *De Artificiali Perspectiva*. Tulli.

PERRONET, Jean-Rodolphe. 1788. *Ses Oeuvres*. Paris: Didot.

PETRIE, W. M. Flinders. 1879. On Metrology and Geometry in Ancient Remains. *Journal of the Antropological Institute of Great Britain and Ireland*, Vol. 8, pp. 106-116.

PEIFFER, Jeanne. 1995. Dürer géométre. Pp. 15-131 in *Géométrie*, by A. Dürer. Paris: Éditions du Seuil.

PLO Y CAMIN, Antonio. 1767. *El Arquitecto Práctico, Civil, Militar, y Agrimensor, dividido en tres libros*. Madrid: Imprenta de Pantaleón Aznar.

RABASA DÍAZ, Enrique. 2000. *Forma y construcción en piedra. De la cantería medieval a la esteorotomía del siglo XIX*. Madrid: Akal.

RANKINE, W. J. M. 1858. *A Manual of Applied Mechanics*. London: Charles Griffin. (3rd ed. 1864.)

RASCH, Jurgen J. 1985. Die Kuppel in der römischen Architektur. *Architectura* **15**: 117-139

RODRÍGUEZ G. DE CEBALLOS, Alfonso. 1983. Entre el manierismo y el barroco, iglesias españolas de planta ovalada. *Goya* **177**: 98-107.

———. 1990. La planta elíptica: De El Escorial al clasicismo español. *Anuario del Departamento de Historia y Teoría del Arte* **2**: 151-172.

ROSIN, P. L. 2001. On Serlio's construction of ovals. *Mathematical Intelligencer* **23**: 58-69.

ROSIN, P. L. and E. TRUCCO. 2005. The amphitheatre construction problem. *Incontro Internazionale di Studi Rileggere L'Antico* (Rome, 13-15 December 2004).

ROSSI, Corina. 2004. *Architecture and Mathematics in Ancient Egypt*. Cambridge: Cambridge University Press.

RUBIÓ, Juan. 1914. Construccions de pedra en sec. *Añuario de la Asociación de Arquitectos de Cataluña* : 35-105.

RUIZ, Hernán el Jovan. ca. 1545. *Libro de arquitectura*. Ms. R.16, Biblioteca de la Escuela Técnica Superior de Arquitectura, Madrid.

SAN NICOLAS, Fray Lorenzo de. S. a. 1989. *Arte y uso de architectura. Primera parte*. 1639. Reprint Valencia: Albatros Ediciones.

SAUVAGE, Martin. 1998. *La brique et sa mise en oeuvre en Mésopotamie: des origines à l=époque Achéménide*. Paris: Centre de Recherche d'Archéologie Orientale.

SERLIO, Sabastiano. 1545. *Il Primo libro d'Architettura di Sebastiano Serlio. . .* Paris: 1545.

————. 1547. *Qvinto libro d'architettvra, nel quale se tratta de diuerse formede tempii . . .* Paris.

————. 1996. *Sebastiano Serlio on Architecture.* Vol. 1: Books I-V of *Tutte l'opere d'architettura et prospetiva*; Vol 2. Books VI-VIII. V. Hart and P. Hicks, eds. New Haven: Yale University Press.

SIMONA, Michea. 2005. Ovals in Borromini's Geometry. Pp. 45-52 in *Mathematics and Culture II. Visual Perfection: Mathematics and Creativity.* M. Emmer, ed. Berlin: Springer.

SMYTH-PINNEY, Julia M. 1989. The Geometries of S. Andrea al Quirinale. *Journal of the Society of Architectural Historians* **48**: 53-65.

VANDELVIRA, Alonso de. 1580. *Exposición y declaración sobre el tratado de cortes de fábricas que escribió Alonso de Valdeelvira por el excelente e insigne architecto y maestro de architectura don Bartolomé de Sombigo y Salcedo, maestro mayor de la Santa Iglesia de Toledo.* Ms. R.10, Biblioteca de la Escuela Técnica Superior de Arquitectura, Madrid.

————. 1977. *Tratado de Arquitectura de Alonso de Vandelvira.* G. Barbé-Coquelin, ed. Albacete: Confederación Española de Cajas de Ahorros.

VIOLLET-LE-DUC, Eugene. 1858. Construction. Pp. 2-208 in vol. 4 of *Dictionnaire raisonnée de l'Architecture Française du XI au XVI siécle.* Paris: A. Morel.

WARD-PERKINS, John Bryan. 1958. Notes on the structure and building methods of early Byzantine Architecture. Pp. 52-104 in *The Great Palace of the Byzantine emperors. Second Report.* D. Talbot Rice, ed. Edinburgh: Walker Trust.

WEST, William Kyer. 1978. Problems in the cultural history of the ellipse. *Techonology and Culture* **19**: 709-712.

WILLIS, Robert. 1843. On the Construction of the Vaults of the Middle Ages. *Transactions of the Royal Institute of British Architects* **1**: 1-69.

WILSON JONES, Mark. 1993. Designing Amphitheatres. *Mitteilungen des deutschen archaologischen Instituts: Römische Abteilung* **100**: 391-442.

ZOCCA, Mario. 1946. *La cupola di S. Giacomo in Augusta e le cupole ellittiche di Roma. (Le cupole di Roma, 4).* Rome: Istituto di Studi Romani.

About the author

Santiago Huerta became an architect in 1981 following study at the School of Architecture of the Polytechnic University of Madrid. He was in professional practice from 1982 to 1989. In 1989 he became Assistant Professor in the School of Architecture of Madrid. He earned a Ph.D. in 1990 with a dissertation entitled "Structural design of arches and vaults in Spain; 1500-1800". Since 1992 he has been Professor of Structural Design at the School of Architecture of Madrid. In 2003 he became President of the Spanish Society of Construction History. From 1992 until the present he has been a consulting engineer for the restoration of many historical constructions, including the Cathedral of Tudela, San Juan de los Reyes in Toledo and the Basílica de los Desampardos among others, as well as some medieval masonry bridges. Since 1983 his research has focused on arches, vaults and domes, and masonry vaulted achitecture in general. He is the author of *Arcos, bóvedas y cúpulas. Geometria y equilibrio en el cálculo tradicional de estructuras de fábrica* (Madrid: Instituto Juan de Herrara, 2004).

Marco Giorgio
Bevilacqua

Dipartimento di Ingegneria
Strutturale
Università di Pisa
Via Diotisalvi, 2
56126 Pisa ITALY
mg.bevilacqua@ing.unipi.it

Keywords: military
architecture, fortress design,
Euclidean geometry, Galileo
Galilei, Francesco di
Giorgio Martini

Research

The Conception of Ramparts in the Sixteenth Century: Architecture, "Mathematics", and Urban Design

Abstract. The discovery of gunpowder and its military applications caused a revolution in the common systems of defence, which had not changed substantially from the Roman period. New methods of laying out urban defences in the second half of the sixteenth century was the product of a continuous response to the evolution of fire arms and their increasing power. The goal of this article is to explain these assertions, analysing in detail the factors that characterized the "science of fortification" in the sixteenth century.

Introduction

Did a relationship really exist between mathematical sciences and the birth and the evolution of the new techniques of defence in the first years of the sixteenth century? And if it existed, what influence did science have on the formulation of the military architecture principles? Was this relationship bilateral, i.e., did the evolution of military architecture also exert a driving influence on experimentation in physics and mathematics?

The answer is not immediate and not certainly simple. So we need to linger over these questions in order to explain if this relationship really existed and what role it played in the wider historical context.

Surely the discovery of gunpowder and its military applications caused a deep change – actually, a revolution – in the common systems of defence, which had not changed substantially from the Roman period. It is also true that the new method of laying out urban defences in the second half of the sixteenth century was the product of a continuous response to the evolution of fire arms and their increasing power.

Moreover, the impulse for the birth of ballistic physics was probably caused in part by military exigencies and, consequently, the improvement of ballistics also determined the mature formalization of the new defensive front.

It is also certain that the new techniques of fortification played an active role in the conception of the Renaissance "ideal town", in which the Euclidean precision of street lines and defensive walls was principally motivated by military requirements.

Gunpowder and the introduction of firearms in war tactics

Some historians attribute the discovery of gunpowder composition to the English friar Roger Bacon (1214-92) at the end of the thirteenth century. Conscious that his discovery could have caused destruction and death, Bacon disguised the ingredients in an anagram: *Sed tamen salis petrae LURU VOPO VIR CAN UTRIET sulphuris, et sic facies tonitrum et coruscationem si scias artificum* (saltpetre ... LURU VOPO VIR CAN UTRIET and sulphur; in this way you will generate thunders and lightning, if you really know the trick) [Luisi 1996, 107]. In the earliest years of the twentieth century, W.L. Hime recognized in the anagram the composition of gunpowder: "R.VII.PART.NOV.CORUL.V.ET. sulphuris", which means "Sed tamen salis petrae recepe VII partes, V novellae coruli, V et sulphuris... ." (...seven parts of saltpetre, five parts of coal and five parts of sulphur...) [Hime 1904].

Apart from other possible interpretations of the anagram, we can deduce that in this first period the approach is still quite alchemic. Subsequently the research was directed towards the improvement of the composition of gunpowder, in order to increase its explosive power, up to the second half of the nineteenth century, when gunpowder was substituted by new stronger explosives, such as nitro-glycerine and dynamite.[1]

Composed of saltpetre (potassium nitrate), coal, and sulphur, gunpowder had already been used in the past for pyrotechnic displays. Only beginning in the fourteenth century was its use applied in Europe for offensive purposes. At the end of the sixteenth century the composition was definitively set forth: six parts saltpetre, one part coal, one part sulphur.[2] These three components were ground into powder, mixed together and then sifted to obtain a more or less fine powder.

If the last two elements were easily available, this was not the case for saltpetre. It was naturally found in form of efflorescence in wet places – such as cellars or caverns – where the action of nitrifying bacteria was possible. Only later were special deposits for nitrification were built for the production and the stocking of rough saltpetre, produced by soil rich in organic waste (the excrement of sheep), diluted with water and then decanted by using boilers.

In contemporary with the improvement of pyrotechnics, the production of firearms started. Even if the first use of firearms is attested in the second half of the fourteenth century,[3] artillery did not assume an important role during battles until to the end of the fifteenth century, when fire arms were still primitive, heavy, and hardly manoeuvrable. Forged in iron, they required a long time to reload and were not very powerful. They were without wheels for being transported and the elevation of the barrel was made earth or crossbars of wood. Projectiles were still in some cases stone and not iron.

A new impetus for the improvement of artillery – and consequently of military architecture – came from the encounter between the French and the Italian armies in 1498, when Charles VIII of France came into Italy to claim his hereditary right to the throne of Naples. His army had superior equipment: cannons were forged in bronze and projectiles in iron. Cannons were more easily transportable by horses, more powerful and required less time for reloading.

The affirmed use of firearms caused an inevitable revolution in the common tactics of war and, consequently, of the ancient techniques of defence: a new age was coming.

New exigencies and first architectonic experimentations

After the introduction of gunpowder at the end of the fourteenth century, medieval fortifications showed themselves to be inadequate for resisting the force of fire arms, when enemies during siege began trying to open a breach in the walls rather than climbing them.

In the second half of the fifteenth century, architects tried to plan new forms of fortifications, in order to improve their resistance to the impact of projectiles. In this period of transition the experimentation of new constructive typologies was principally applied to fortresses and citadels (figs. 1-4). Even if the character of these was still deeply medieval, it is possible to recognize in them some innovative typological elements.

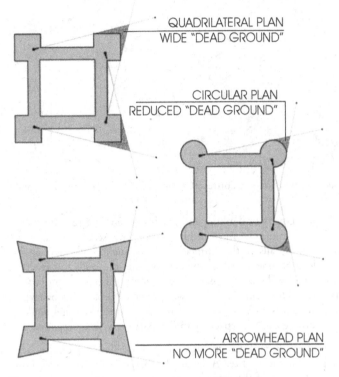

QUADRILATERAL PLAN
WIDE "DEAD GROUND"

CIRCULAR PLAN
REDUCED "DEAD GROUND"

ARROWHEAD PLAN
NO MORE "DEAD GROUND"

Fig. 1. Types of towers and "dead ground"

Fig. 2. View of Volterra Fortress (Francione, 1472)

Fig. 3. Harquebus slit, Volterra, fifteenth century

Fig. 4. Plan of Volterra Fortress (G. Panazzi, 1755). Courtesy of the Biblioteca Guarnacci, Volterra

First of all, the recovery of flanking and grazing defensive fire as a generating principle is the main point of this new way of planning.[4] The search for a regular planimetric conformation of the walls is the immediate consequence of this principle, in order to minimize the "dead ground", that is, the area outside the walls not reachable by defensive shots. The corners of walls were reinforced by cylindrical towers, projecting outward from the boundary walls. The cylindrical form was preferred because its behaving as an arch was more resistant to shots. The circular form also optimised the flanking defence and minimized the "dead ground", especially in comparison with square towers.

The height of walls and towers was limited; a high tower, in fact, represented a dangerous target for enemy fire. A great number of slits for small bombards and the harquebus (a predecessor of the musket) were opened along the perimeter of the fortress.[5] Towers were larger in diameter than medieval ones, to allow for placing firearms and accommodating their recoil. The boundary walls had a great scarped revetments, a concept called *mura fuggitive* (literally, fugitive walls), that is, the property of sloping walls to deflect cannonballs.

All around the perimeter, the moat was deeper and larger, in order to keep enemy posts as far as possible from walls. All the same, the crowning of towers were still typically medieval, with brackets, merlons and trapdoors, which show that climbing walls persisted in siege tactics.[6]

The new attention paid to geometrical simplicity in plan represented the modern character of military architecture. From cylindrical towers to arrowhead bastions, the step was short.

Some historians attribute the paternity of the bastion to Francesco di Giorgio Martini, after his famous drawing from the *Codice Magliabechiano* (fig. 5).

In this period of experimentation, Francesco di Giorgio was surely the most innovative architect. He was really the first to reintroduce flanking as the most important principle in fortress project.

Fig. 5. Francesco di Giorgio Martini. La rocca di San Leo

Fig. 6. Giuliano da Sangallo, detail of the *fianco ritirato con orecchione* in one of the bastions of the
Fortezza Nuova in Pisa

But if Francesco di Giorgio understood that the arrowhead form of towers was more effective, we must attribute to Giuliano and Antonio da Sangallo the mature definition of the bastion's formal parameters and the modern way of fortifications (fig. 6). In particular, they invented a special form of bastion, which became an indispensable element in planning fortifications in the centuries that followed: the bastion with orillions, *fianco ritirato con orecchione*, i.e., the flank of bastion where the embrasure was hidden by a curved projection that defended it from enemy's cannonade.

Euclidian geometry in the treatises of military architecture

At the end of the fifteenth century, architects began conceiving new methods for the design of urban defences. If until that time the figure of the architect had retained the characterization given to him by Vitruvius – that is, of a man of universal learning – in consequence of the technological revolution induced by firearms, a slow process of specialization started. The figure of military architect was born as a specialist in military problems.

The desire to set forth their own experiences induced many of these architects to write specialized treatises on fortifications, war tactics and ballistic theories.[7] The cultural discussion took place in an international context, all over Europe.

The importance of Euclidian geometry clearly emerged from these works, whether written by architects or by soldiers. It is not by chance, for example, that in the frontispiece of his treatise of 1537, *La Nova Scientia*, Niccolò Tartaglia wrote, *Nemo Huc Geometriae expers ingrediat* (in effect, let him who will understand these theories have a thorough knowledge of (Euclidian) Geometry). His work, which was the first attempt at a mathematical treatment of the movement of projectiles, was really innovative and became a point of reference for all the other writers. In particular, his study tried to geometrize the trajectory of projectiles on the basis of the force of lauch and of the inclination of cannon barrels (fig. 7).

With Jacomo Lanteri (1557), moreover, military architecture was clearly considered not as a design practice, but very much a part of the mathematical sciences: he declared that the finality of his treatise is "to talk about the way of planning fortresses using Euclidian principles".[8]

At the end of the sixteenth century, while he was professor of Mathematics at Padua University, Galileo Galilei wrote two treatises. In his *Trattato di architettura militare* (1593) the approach to the problem is strongly geometrical and mathematical. Thus, before starting to write about architecture, he reserved the first pages for the explanation of graphical methods for solving geometrical problems, necessary for planning ramparts: how to trace perpendicular lines, for example, how to divide a corner in equal parts, or to draw regular polygons.

With Galileo, the relationship between military architecture and mathematics was universally recognized. That's the reason why we choose to refer to his rules in describing, even if briefly, the formal principles of "modern" way of planning fortifications (fig. 8).

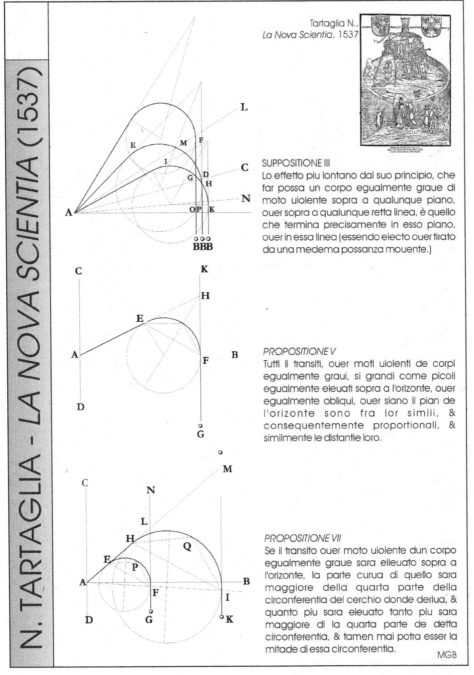

Tartaglia N.,
La Nova Scientia, 1537

SUPPOSITIONE III

Lo effetto piu lontano dal suo principio, che far possa un corpo egualmente graue di moto uiolente sopra a qualunque piano, ouer sopra a qualunque retta linea, è quello che termina precisamente in esso piano, ouer in essa linea (essendo eiecto ouer tirato da una medema possanza mouente.)

PROPOSITIONE V

Tutti li transiti, ouer moti uiolenti de corpi egualmente graui, si grandi come picoli egualmente eleuati sopra a l'orizonte, ouer egualmente obliqui, ouer siano il pian de l'orizonte sono fra lor simili, & consequentemente proportionali, & similmente le distantie loro.

PROPOSITIONE VII

Se il transito ouer moto uiolente dun corpo egualmente graue sara elleuato sopra a l'orizonte, la parte curua di quello sara maggiore della quarta parte della circonferentia del cerchio donde deriua, & quanto piu sara eleuato tanto piu sara maggiore di la quarta parte de detta circonferentia, & tamen mai potra esser la mitade di essa circonferentia.

MGB

Fig. 7.

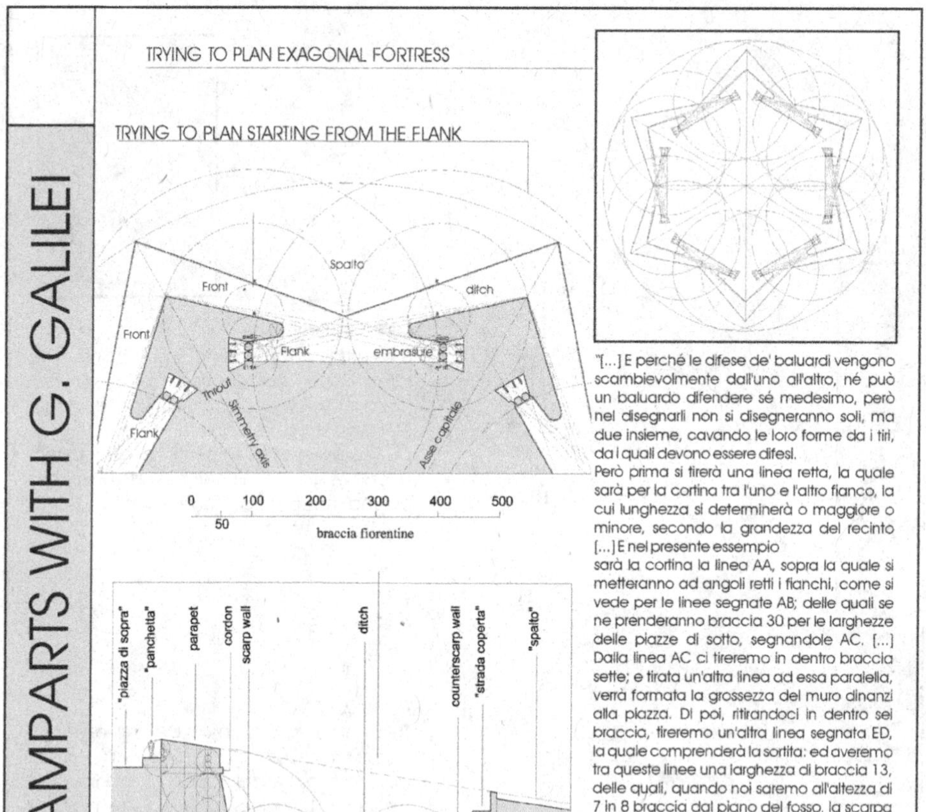

TRYING TO PLAN RAMPARTS WITH G. GALILEI

TRYING TO PLAN EXAGONAL FORTRESS

TRYING TO PLAN STARTING FROM THE FLANK

Spalto

Front — ditch

Front

Flank — embrasure

Throat — Symmetry axis

Flank — Asse capitale

0   100   200   300   400   500
50

braccia fiorentine

"piazza di sopra" · "panchetta" · parapet · cordon · scarp wall · ditch · counterscarp wall · "strada coperta" · "spalto"

scarp wall
wall
counterfort
embankment

0   5   10   20   30   40   50

Braccia Fiorentine

DRAWING SCARPS, DITCHES AND COUNTERSCARPS

MGB

"[...] E perché le difese de' baluardi vengono scambievolmente dall'uno all'altro, né può un baluardo difendere sé medesimo, però nel disegnarli non si disegneranno soli, ma due insieme, cavando le loro forme da i tiri, da i quali devono essere difesi.

Però prima si tirerà una linea retta, la quale sarà per la cortina tra l'uno e l'altro fianco, la cui lunghezza si determinerà o maggiore o minore, secondo la grandezza del recinto [...] E nel presente essempio

sarà la cortina la linea AA, sopra la quale si metteranno ad angoli retti i fianchi, come si vede per le linee segnate AB; delle quali se ne prenderanno braccia 30 per le larghezze delle piazze di sotto, segnandole AC. [...] Dalla linea AC ci tireremo in dentro braccia sette; e tirata un'altra linea ad essa parallela, verrà formata la grossezza del muro dinanzi alla piazza. Di poi, ritirandoci in dentro sei braccia, tireremo un'altra linea segnata ED, la quale comprenderà la sortita: ed averemo tra queste linee una larghezza di braccia 13, delle quali, quando noi saremo all'altezza di 7 in 8 braccia dal piano del fosso, la scarpa ne avrà consumato braccia 1½, tal che resteranno braccia 11½, sendo scemato dalla parte di fuori lo spazio sino alla linea FG. Pigliando dunque il mezo tra DE ed FG, vi tireremo la linea HI, sopra la quale, cominciando dalla cortina A, misureremo due braccia per la prima cannoniera: doppo, pigliando con il compasso la misura di braccia 5 e quarti 3, segneremo un cerchio, che avrà di diametro braccia 11½; doppo il quale si lascieranno due altre braccia per la larghezza della seconda cannoniera: doppo la quale disegneremo un altro cerchio con il medesimo diametro, e doppo esso due altre braccia per la terza cannoniera: e così delle 30 braccia ne averemo consumate 29, cioè 23 per li due cerchi, che ci rappresentano due merloni, e 6 per le tre cannoniere; e quel braccio che avanza, servirà per risalto o spalietta. [...] La lunghezza della piazza non si farà meno di 40 braccia, facendola pendere un poco verso la fossa, acciò che anco dall'estremità di dentro possa fare effetto; e perciò si faranno le cannoniere senza soglia o scaletto.
From Galilei G., *Trattato di Fortificazione*, 1593

Fig. 8.

Modern way of planning fortifications. Geometry and structure

The formal symmetry, proportion and regularity of new fortifications surely reflected the Renaissance culture and aesthetic canons. Otherwise, their conformation to geometricy was justified by precise military reasons. The bastion was the most important structure of this complex system, where each element was deeply connected to the others by principles of reciprocal defence.

In its simplest configuration, this system was principally formed by two adjacent ramparts and the walls between them (fig. 5). The profile of each element was generated by precise method of planning, based on the trajectories of projectiles, assuring the defence of walls between ramparts and the reciprocal defence between ramparts, and cancelling any "dead ground".

The rampart showed a great planimetric symmetry, linked to the bisecting axis – called *asse capitale* – of the corner it was built to defend. It had five sides: two fronts, two lateral flanks and the ideal internal one called the *gola* (literally, throat).

Generally, firearms were put in two levels of fire. A first level was constituted by the *piazza di sotto* (lower square), one for each flank, sometimes in the form of one or two levels of casemates, symmetrically opened in the flanks, near the walls, with two embrasures each for the grazing defence of the walls. A superior level was represented by the *piazza di sopra* (upper square), a large, open-air square, with embrasures opened along all the perimeter.

Inside the body of the rampart, at the level of foundations, a vaulted gallery ran all along the perimeter, with air vents for the ejection of the smoke from the firearms.

This gallery had a double function: like a real casemate along the flanks, it had many harquebus slits; along the fronts it permitted inspection of the perimeter (countermine gallery), against possible enemy mines.

Outside, a large moat ran along all the perimeter; its form responded to the same methods of planning. Over the ditch was the *spalto*, a flat space wider than a mile, without trees, building or any other elements that could offer repair to enemies.

The boundary was made of brick or stone walls, reinforced inside by counterforts. These were rectangular in plan or, as suggested by Galilei, pentagonal, in order to maintain the earthen ramparts in case of breach.

Foundations were usually built of fired bricks, hard bricks or hard stone, because of their excellent structural resistance and resistance to humidity.

For external wall surfaces, on the other hand, green (less baked) bricks or sandstone (*pietra morta*) were used, because they better absorbed cannonball shots. Where it was damp, fired bricks were preferred only for the external wall surface, or generally on the inclined surface of parapet, being more resistant to rain and natural agents.

The filling between counterforts was made with tamped clay. The earth inside the ramparts was generally made stable with branches of chestnut, oak or of any other water-resistant quality of wood.

In the second half of the sixteenth century, this first model of bastion *all'italiana* (in the Italian way) showed its vulnerability. In fact, with the evolution of fire-arm technology and pyrotechnics, using brick or stone became dangerous, because, when such walls were hit by projectiles the fragments that flew into the air were almost as dangerous as the projectiles themselves.

Moreover, the practice of digging underground tunnels to reach the bastions[9] was more frequently used. So in order to defend ramparts from enemy attacks and mines, new defence structures were built in and outside the ditch (*mezzelune, falsebraghe,* etc...).

By the end of the sixteenth century, the Italian model of ramparts was definitely substituted by French or Flemish ones, made completely of earth, limiting walls to the basement.[10]

From ideal to real town. A case study: Leghorn

Between the end of the fourteenth century and the first half of the fifteenth century, very significant changes occurred. The protectionist economy based on medieval corporations gradually ended. International trade improved in consequence of the discovery of the New World. Moreover, a process of centralization of territories and powers gradually occurred, with the birth of national and regional monarchies.

In this context, medieval towns were subjected to rigid control by the central government. The medieval urban structure, with its "labyrinth" made of narrow streets, surely did not respond to this necessity.[11]

The consequent regularization of the shapes of towns was certainly produced by the new Renaissance culture, where nature was subjected to geometry and the architectural language was definitely codified in accordance with the classical principles of symmetry and proportion.

However, military exigencies certainly played an important role in defining the Renaissance model of the star-shaped town. Large streets, in fact, permitted the rapid deployment of men and weapons during sieges. The wide central square, symbol of political power, both civil and religious, was also used as parade-ground in war periods. Finally, the star-shaped boundary walls, with ramparts and external defensive structures, clearly showed the military character of the Renaissance town. The removal of the defence-line as far as possible from town centre sacrificed outer-wall suburbs and lands to the exigencies of war, irremediably changing the ancient relationship between town and country-side.

Actual examples, such as Palmanova, ideal ones, such as *Sforzinda* (Filarete, 1460) or utopian ones, such as *Christianopolis* (J.V. Andreae, 1619) showed that geometry and regularity, apart from respecting principles of formal perfection, responded to precise functional exigencies, assuring the town's defensive efficiency.

As an example of the new urban design related to military necessities, we choose to introduce the first urban plan of Leghorn, drawn by Bernardo Buontalenti in 1576 (fig. 9).

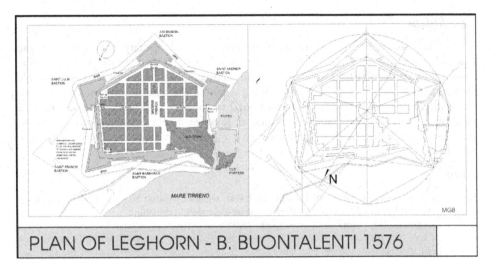

PLAN OF LEGHORN - B. BUONTALENTI 1576

Fig. 9.

The plan had a mostly regular pentagon shape. The irregularities derived from the presence of the western old town, with the fortress drawn by Giuliano da Sangallo near the port. In the new plan, the old fortress was placed in one of corners of the defensive boundary and the old town was fortified with ramparts.

Along the perimeter, Buontalenti projected five new bastions: Ascension's Bastion at the southeast, Santa Barbara's Platform[12] at the northwest, St. Andrew's Bastion at the southwest, St. Julia's Bastion at the east and St. Francis's Bastion at the north.

Defence lines measured no longer than 750 ells. The fronts of bastions were unusually designed on defence lines starting from the corner of perimeters and not from the corner between flanks and walls.

In the middle of each side of the perimeter, Buontalenti built some secondary defensive structures, called *cavalieri*, (knights) for shooting outside the city and defending the nearest ramparts from the middle of the curtain.

The scheme of orthogonal streets permitted soldiers to cross the city quickly. The principal axis was represented by the main street, 14 ells wide. It had an inclination of 30° with respect to the North, in order to optimise the effects of sun and wind.

The main square in the centre, with the Cathedral, was really a parade ground, directly connected with the trapezoidal service square behind the *gola* of Ascension's Bastion.

The "Via Giulia", the other main street, orthogonal to the one mentioned above, served commercial and military purposes, connecting the two most important city gates – Porta a Mare and Porta a Terra – placed near the flanks of bastions and efficaciously defended by their embrasures.

A circumvallation 20 ells wide ran inside along all the defensive perimeter and permitted troops to reach ramparts by a complex system of rhomboidal-shaped ramps.

Conclusions

So, did a relationship exist between mathematics and military architecture in the sixteenth century? If so, Euclid surely presided over it. And, more generally, if the scientific approach is based on analysing the relationship between causes and effects of any possible phenomena, in the "modern" way of planning defences, the regular and geometrical shape of fortifications was surely conceived to respond to concrete exigencies. That was really a modern attitude, where shape and function lived in perfect equilibrium, so deeply characterizing the face of cities and their civil organization from the sixteenth century onwards.

Notes

1. Nitro-glycerine was first obtained by the Italian scientist Ascanio Sobrero in 1847, and then perfected by Alfred Nobel in 1866 with the introduction of dynamite.
2. The present-day composition is: saltpetre 74.64%, sulphur 11.85%, coal 13.51%.
3. In Italy, the first use of bombards is attested in 1331 in the siege of Cividale. Firearms have been illustrated in codices ever since 1325. A document about *Provvisione* of the Florence Republic (dated 11 February 1326) is perhaps the first to talk about firearms.
4. The flanking principle of defence was already present in the Roman age. According to Vitruvius, in fact, summarizing Greek and oriental details, towers must jut out over the walls in order to strike enemies sideways, and the longest possible distance between towers was based on the range of arch arrows.
5. The typical harquebus slit had, outside, the shape of an upside-down keyhole and permitted a range of fire 20° wide. The hole was for lodging the barrel of the harquebus, while the vertical slit was for aiming the shot. Inside, it had a simple splay, showing its direct derivation from the loophole.
6. One of the most representative examples of this type of fortifications is the Medici Fortress in Volterra, planned by Francesco di Giovanni di Matteo Francione in the 1472. The fortress is constituted by two parts joined together by two long walls. The eastern one, called *Mastio* or *Maschio* (the Male), has a quadrilateral perimeter with cylindrical towers at the corners and a big cylindrical tower in the centre, overlooking the town centre and built against possible rebellions of its citizens. The western one has a semi-cylindrical tower (called the Female) and a great arrowhead fortification toward the valley, a sort of big polygonal armed square – quite a bastion – looking outside the city.
7. Some of the most important treatises in military architecture:
    Roberto Valturio, *De re militare*, 1472.
    Antonio Corazzano, *De re militaria*, 1493.
    Giovan Battista Della Valle di Venafro, *Vallo*, 1521.
    Niccolò Machiavelli, *Arte della Guerra*, 1521.
    Albrecht Durer, *Etliche underricht, zu befestigung der Stett, Schloss, un flecken*, 1527.
    Nicolò Tartaglia, *Nova Scientia*, 1537. *Quesiti et invenzioni diverse*, 1538.
    Giovan Battista Belluzzi, 1545-98.
    Pietro Cataneo, 1567.
    Giovan Battista Zanchi, 1554.
    Francesco De Marchi, *Architettura militare*, 1565.
    Giacomo Lanteri, *Del modo di fare le fortificationi di terra intorno álle città, & alle castella per fortificarle. Et di fare cosi i forti in campagna per gli alloggiamenti de gli esserciti; come anco per andar sotto ad una terra, et di fare i ripari nelle batterie*,1557.
    Francesco De Marchi, 1599.
    G. Maggi-I. Castriotto, *Della fortificatione della città*, 1564.
    Girolamo Cataneo, *Libro Nuovo di fortificare, offendere et difendere con alloggiamenti campali*, 1567.

Galazzo Alghisi, 1570.

Domenico Mora, *il Soldato*, 1570.

Gabriello Busca, *Della espugnatione delle fortezze*, 1585.

Galileo Galilei, *Breve istruzione all'architettura militare,* 1592-93. *Trattato di fortificazione,* 1593. *Del compasso geometrico e militare,* 1606.

Bonaiuto Lorini, *Le fortificationi,* 1597.

8.   *A ragionar del modo di disegnare le fortezze secondo Euclide* [Lanteri G., *Del modo di fare le fortificationi di terra intorno alle città*, 1557].

9.   A special corps of sappers, digging a gallery underground, reached ramparts and made them explode with gunpowder.

10.   Galileo showed in his treatise the way to compress earth by using wooden chains and to protect rampart's external surfaces by turfing or using sun-dried clay blocks. Even if the Flemish models better responded to the new techniques of siege, they required more upkeep and were more subject to deterioration.

11.   In 1475 Ferrante d'Aragona, king of Naples, defined narrow streets a "danger for the State".

12.   Platforms were similar to bastions but they differed in being placed in the middle of defensive perimeter and not in the corner.

References

BOCCIA, L. G., J.A. GODOY. 1986. Armeria I and II. Exhibit catalogue (2 vols), Museo Poldi Pedazzoli. Milan: Electa.

CAGIANELLI, F. 2001. *Livorno: la costruzione di una immagine.* Cisinello Balsamo: Silvana Ed.

CASSI RAMELLI, A. 1996. *Dalle caverne ai rifugi blindati.* Bari: Mario Adda Ed.

CECCHETTI, R. 2006. *Forma della città e gruppi sociali.* Pisa: Plus Ed.

FARA, A. 1995. *Bernardo Buontalenti.* Milan: Electa.

———. 1993. *La città da guerra.* Turin: Einaudi.

FIORE F.P., M. TAFURI (eds). 1994. *Francesco di Giorgio Martini architetto.* Milan: Electa.

GALILEI G. 1980. *Trattato di fortificazione.* Series *I Classici delle Scienza,* F. Brunetti, ed. Turin: UTET.

HIME, W.L. 1904. *Gunpowder and Ammunition, their Origin and Progress .* London.

LUISI, R. 1996. *Scudi di pietra. I castelli e l'arte della guerra tra Medioevo e Rinascimento.* Bari: Laterza.

MARTINI, Francesco di Giorgio. 1967. Architettura civile e militare, in *Trattati di architettura, ingegneria e arte militare,* vol. 2. C. Maltese, ed. Milan.

MUMFORD, L. 2002. *La città nella storia.* Milan: Bompiani.

SEVERINI, G. 1970. *Architetture militari di Giuliano da Sangallo.* Pisa.

———. 1994. *Progetto e disegno nei trattati di architettura militare del '500.* Pisa: Pacini Editore.

TARTAGLIA, N. 1984. *La nova scientia.* Bologna: Arnaldo Forni Editore.

ZDEKAUER, L. (ed.). 1979. *Il taccuino di Giuliano da San Gallo.* Siena: Arnaldo Forni Editore.

About the author

Marco Giorgio Bevilacqua earned his degree with honors in 2003 in Civil-Structural Engineering at the University of Pisa. Since 2004 he has been working on his doctorate in "Science and Techniques for Civil Engineering at the same university. His thesis is entitled "La fortificazione della città di Pisa nel XVI secolo. Il primo fronte bastionato". He is part of various national and international research projects on the subjects of the designing and surveying of architecture. He collaborates in teaching projects for architectural design in the first years of the degree program in structural engineering-architecture at the University of Pisa. Beginning this year, he has a contract to teach a laboratory for CAD applications. Only in his spare time is he a "real" engineer.

Dirk Van de Vijver

Sint-Gummarusstraat 25
B-2060 Antwerp
Dirk.vandeVijver@let.uu.nl

Keywords: structural
mechanics, Belgian Royal
Academy, De Nieuport,
arches, vaults, beams,
structural analysis

Research

Tentare licet. *The Theresian Academy's Question on the Theory of Beams of 1783*

Abstract. The four answers to the prize question of the Brussels' Academy of 1783 on the development of a theory of beams demonstrates that the modeling and mathematical mastering of the problem remained for a long time limited to a very small circle of men. With *savants* such as the Viscount de Nieuport (1746-1827), who worked on the calculus of vaults and who formulated this question on beams, the Academy appears such a privileged milieu. In Belgium, it would at take least until the creation of the polytechnic school at the State University of Ghent in 1835 for this approach to be diffused in a systematic and institutionally way among (an elite of) construction professionals.

Introduction

The exact sciences' section of the *Theresian Academy* of the Austrian Netherlands proposed the formulation of a theory for beams as the subject of the annual competitions of 1783. The four answers to this question which this institution received are kept in the Old Archives of the Belgian Royal Academy[1] and constitute one of the few written sources on the subject, valuable information on an interesting moment in the history of building mechanics in Belgium, which remains to be written.[2]

Although the competition was not a success – none of the four manuscript papers received a prize or even an honorable mention and, afterwards, the subject was abandoned –, an analysis of these papers offers valuable insight into the state of the art on the problem in the Austrian Netherlands. The difficulties of these authors with the subject reveal the novelty of the approach represented by a member of this academy, Viscount de Nieuport, who had worked and published before on the calculus of vaulted structures and who invented the prize question discussed here.

The Theresian Academy, de Nieuport and the "scientification" of building construction

The "learned society" in the Austrian Netherlands, equivalent to the French *Académie des Sciences* and to the English *Royal Society* was instituted by patent letter of Maria Theresia in 1769 as *Société littéraire*, and transformed, on 16 December 1772, into the *Académie impériale et royale des sciences et belles-lettres de Bruxelles*.[3] Like the French *Académie des Sciences* this institution was patronized by the central government (of the Austrian Netherlands), and counted among its members only scientists, *savants*. From its foundation, the academy consisted of two sections: the *classe historique* – a department with historians – and the *classe physique* – a department with mathematicians, physicists, military engineers, geologists, chemists and physicians. Both sections reflect the double mission of the Academy: to write a national history and to apply the sciences in order to enhance the economical prosperity of the Austrian Netherlands. From the beginning it was clear that the institution stood in service of the nation: "the first object without doubt that our newly born society must look for is being useful to the country", wrote Abbé Needham in 1769.[4]

The small group of members of the Academy met regularly to discuss their work. In these sessions as well prize questions were chosen, and papers answering those questions were studied, evaluated, and eventually rewarded with a prize and with their publication in the Academy's publication series. The system of prize questions was an affirmed strategy for engaging the learned population in contributing to questions judged of national interest by the Academy and to enhance the emulation of the competitors.

In the Austrian Netherlands, the Theresian Academy was the only scientific institution which was actively involved in the theory of hydraulics, building mechanics, fireproof constructions and building materials, and which published on those subjects [Van de Vijver 2003a, 108-119 ("De Theresiaanse academie en het bouwen")]. (The curriculum at the Old Leuven University remained limited to basic courses; some of them, such as the ones on architecture, on mechanics and on surveying weren't even printed [Van de Vijver 2003a, 102-107 ("Architectuur aan de universiteit")].) Members of the Academy, such as the Englishman Theodore-Augustin Mann (1735-1809) [Van de Vijver 2003a, 111-117, cat. n. 39 ("Een overheidsonderzoek naar brandveilige constructies")], who published on hydraulics and on fire proof constructions, operated within a broad European context. In the Southern Low Countries, they belong to the principal representatives of an evolution towards the 'scientification' of building construction in the last third of the eighteenth century.

Another member of the Theresian Academy, Charles-François-Ferdinand le Prud'homme d'Hailly, Viscount de Nieuport (1746-1827), lieutenant of the corps of military engineers of the Austrian Netherlands, was the first mathematician of the country to model vaults as consisting of infinitesimally small parts and to apply the calculus of differentials to them.[5] Like Claude-Antoine Couplet (1729), Georg Wolfgang Krafft (1754-1755) and Abbé Charles Bossut (1770), de Nieuport tried to apply in his *Essay analytique sur la méchanique des voûtes* (presented at the Academy on 18 May 1778 and published in 1780), the theoretical model proposed by Jacob Bernoulli [De Nieuport 1780; Radelet-de Grave 1995]. In a second study, entitled *Mémoire sur la propriété prétendue des voûtes en chaînettes* (presented at the Academy on 6 November 1780 and published in the *Mémoires de l'Académie impériale et royale des sciences et des belles-lettres de Bruxelles* in 1783), he commented on a paper by Krafft that contradicted a theoretical fundament of Jacob Bernouilli [De Nieuport 1783; Radelet-de Grave 1995]. The Brussels Academy preferred these memoirs in applied mathematics over his more purely mathematical subjects: the former were considered to be more useful to the general public. In his report on the *Essay analytique sur la mécanique des voûtes*, Abbé Mann states: "This important paper seems in every way well done, curious and interesting. I don't hesitate to say that it merits a place in the second volume of our *Mémoires*, in preference to two other papers of the same author, which treat subjects less useful than this one".[6] Abbé Chevalier judged the publication of the latter paper important "for the public good" (*pour l'utilité publique*) [Mailly 1883, II: 110 (session of 12 October 1780)]. However, although the subject of vault construction came from ordinary life, Nieuport's approach and the mathematical methods applied by him were certainly not. As Patricia Radelet de Grave argues, de

Nieuport's approach shows clearly the author's obsession for the mathematical aspects of the problem.[7]

The prize question of a theory on beams

The Viscount de Nieuport's proposition of the mechanical behavior of beams as a subject for an academy prize question was discussed and agreed upon in the Theresian Academy's session of 19 October 1781.[8] The question, distributed in French and Dutch in the principal newspapers, was formulated as follows:

> Develop a theory on beams which rest with both ends upon points of support, considering them in the hypothesis most conform to nature, that is, as a collection of weighting, extensible and elastic fibers, united among each other over the whole length. Deduct from this consideration the cause and the place of rupture in the different cases of solicitation by weights, and, consequently, determine the best use of hanging trusses.[9]

The research question is formulated with precision and detail. It specifies the way of modeling the material (wood) – "in the hypothesis most conform to nature, that is, as a collection of weighting, extensible and elastic fibers, united among each other over the whole length" – and fixes the geometrical side-conditions of the problem: a beam which "rest[s] with both ends upon a point of support". Furthermore, the general question on the development of a theory on beams is defined as a response to two fundamental problems: the cause and place of rupture in different cases of solicitation; and an applied, correlated problem: how to strengthen the structure by secondary means (*liens pendants*) – an applied, practical, real-life aspect absent in de Nieuport's earlier papers on vaults.

With the specification of the geometrical side-condition, the Academy got rid of the statically more complicated, but more realistic situation of a beam wedged in a wall – turning at the same time a hyperstatic problem into a static one. The hypothesis of considering the material as consisting of elastic fibers refers to the modeling of solids by Jacob I Bernouilli (1654-1705) [Bernouilli 1705]; it makes it not only possible to describe the fiber/cord under solicitation mathematically with an "elastic curve", but also eliminated the complexity of the real composition of a wooden beam – a natural product which not only depends on the species, growing conditions and incidental deficiencies, but as well on the manufacturing process of a stem of a tree into a beam – by replacing it with an "ideal material". These two elements reduce the complex reality into an abstract model disposed to a mathematical solution. The proposed modeling of the problem permits the assumption that de Nieuport expected a response in the line of his work on vaults, in which mathematical tools helped to understand the mechanical behavior of a structure.

The question of the cause and place of rupture in different cases of solicitation refers of course to Galileo's treatment of the problem [Galileo 1638]. The absence of words and concepts such as "strength of wood" (*force des bois*) or "experiments" make it clear that de Nieuport and his colleagues of the Academy didn't want to orient the question towards the experimental study of the resistance or strength of wood, a field in which many scholars became particularly active in the eighteenth century (for example, Edme Mariotte, Bernard Forest de Belidor, Pieter van Musschenbroek, Antoine Parent, Duhamel de Monceau, and Buffon).[10]

The conclusion of the commissioners after the lecture of the answers received –"neither to present any prize, nor to give any honorable mention and to abandon the question"[11] – already suggests that they did not receive an answer in the direction suggested by the line of questioning, and that we should not expect to find an innovative mathematical or mechanical approach to the problem, nor an answer which falls within the canon of historical solutions to this mechanical problem.

The Flemish answers

Three of the four *mémoires*, or papers, received, were written in Dutch and are rather brief. In fact, as they usually were read aloud in the Academy's sessions, the length of the contributions was limited to a text of about half an hour. The only paper in French stands a bit outside this group, not only because of its length (the text numbers forty-one pages and eight full-page illustrations), but above all because of its "learned" content. In order to enhance an unbiased judgment, it was standard practice to send in the answers anonymously, only marked by a motto. As the prize was not distributed, all authors of the papers remain unidentified. We shall identify the anonymous papers by their motto and actual archive number.

Of the four papers, one paper can already directly be excluded from further analysis, as it doesn't give an answer to the question. In fact, the author of paper ARB AA 215 with the motto *Alle deeze t'saemen-gevoegde/ is hart met hart* ("All these united, heart with heart"), transposed the question to a whole different field of reference (religion and death penalty), which falls totally outside our topic.[12]

Of the other two Flemish papers, the hypothesis of a homogenous beam doesn't seem to have been an obvious modeling of the material at all. For the author of memoir ARB AA 213, with the motto *De oordeelen der menschen worden heel verschillig gevonden, op eene merkweerdige zaeke* (Men's judgments are very different in a curious matter), the hypothesis of an "ideal" beam without deficiencies (*weeren ofte quaede plekken)*, such as the one described in the academy question, isn't an obvious assumption at all. Four out of ten paragraphs enumerate the possible deficiencies of "real beams": they differ regarding the part of the tree they come from (§ 2); heartwood will split (§ 3); trees which are not straight give beams with cut fibers which compromise the strength; also beams which are larger on one side have usually cut fibers (§ 4); bad spots in wood (*groote were of quaede plekken*) will split or fall out; it is preferable to use hard wood such as oak and to avoid *morise hout*, which won't survive insect attack (§ 5). Unlike the author of paper 213, which after his objection goes one with the suggested hypothesis, memoir ARB AA 216 with the motto *Den Zang verheugd/ In eer en deugt/ Hunne Zoetigheijt/ Maeckt ons verbleijt* (Singing rejoices/ In Honor and Virtue/ Their Sweetness/ Makes us happy), never accepted the proposed modeling of the wood.[13] As a tree is smaller upwards and thicker downwards, the author of the latter paper argues that a straight wooden beam without faults will crack a bit out of the middle. (Note that this even makes the problem asymmetrical!) This "bit" is different for each beam/tree, he states. He concludes that there is no fixed rule for all beams, because the differences of the wood make that impossible: it's different each time.[14] The author indicates the side of the crack a "bit" near the middle: the one which corresponds to the former upper side of the tree. The argument is that the wood there is "sweeter of fiber" (*souter van draet ofte vesels*) – less weight makes it more fragile. In fact, the author of paper 216 doesn't seem to accept the implicit suggestion imbedded in the question that in the proposed model the wood consists of fibers of a uniform diameter.

For the author of paper 213, the place of failure for symmetrical loading conditions is in the middle of the beam "in the union or joints of the fibers to which place the weight always takes its refuge",[15] or in other loading conditions at the place of the weight/solicitation. For him the reason of failure of a pure wooden beam without deficiencies is an "internal fracture of the fibers" (*inwendige breuke, splinteren ofte veselen*).

The same author considers the use of *hangende banden* (hanging trusses) as a polyvalent solution, which permits for instance to use core wood, and prevent it from splitting; also for beams with split fibers, the bond is used to hold wood together. He proposes one or three "bonds", the strongest in the middle, the others at the ends (§ 8). To prevent a profusion of bonds, the author proposes combining an iron plate under the beam (as well as at the supports) with a bond in the middle, complemented as necessary with two supplementary bonds at both ends (§ 9). This method is also advised for a beam composed of different parts of wood (this is wood with cut fibers) to prevent cracking, and in the many cases where wood fibers are not continuous (bent trees, sick spots or wrongly cut wood) (§ 10). The author of *mémoire* AA 216 says that the best way to strengthen the wood, is to place a support in the middle (or a bit out of the middle) – a solution only adopted in bonded warehouses. Subsequently, the author enumerates a series of alternative ways of support: (a) to add a second beam on top of the first, and to join them with a screw, using a small piece of wood under the original first beam; (b) to attach a piece of wood under the original beam by screwing, using an iron device, or by nailing; (c) to use pieces of wood neither over nor under the beam, but at the sides, and to join them with pieces of iron; (d) in cases of strong loads upon the beam, use a masonry arch to evacuate the loads: the beam can be hung by iron tires to the masonry arch; (e) the arch can be replaced by a wooden triangle; (f) a short piece of wood can be fixed on top of the beam; or, finally, (g) an iron plate can be put under the beam.

Tentare licet

The French paper, entitled *Tentare licet* ("One may try"), ARB AA 214, is the most extensive paper and the only illustrated one. It is also unique in mentioning some authors, such as Pieter van Musschenbroek (1692-1761) and Jean-Antoine Nollet (1700-1770).[16] This already suggests that here again we shouldn't expect a mathematical answer: the paper is more situated in an experimental approach of physics where tests determine the strength of materials, and the resistance of wood.

The *mémoire* is structured in chapters and principles. The first chapter goes into the so-called "intrinsic reactive force" of the fibers, the second deals with the causes of rupture and the places where they occur, the third and final chapter explains the ways to strengthen the beam; in an appendix the opinions of a carpenter and an architect of the region of Reims are given, illustrated with some case studies derived from their practices.

In the first chapter the author introduces the intrinsic reactive force (*force intrinseque réactive*) of the longitudinal fibers: a force which reacts against the perpendicular weight which makes the fibers bow and break. He states (a) that this force is in direct relation to the horizontal surface of the supporting ends; (b) that it is inversely related to the distance of the supporting ends (the length of the beam between the supports); (c) that it is more directly related to the height than to the width of the beam; and, finally, (d, related to a), that without support at the ends, there is no reactive force. To illustrate the latest point he quotes an experiment taken out of the *Récréations mathématiques et physiques* by Jacques Ozanam [1778]. In this chapter, the author especially exploits the fiber-model of wood to warn that a fiber is only effective when it remains intact and when it is supported at the ends, condemning general habits, such as the practice of building craftsmen who destroy a part of the beam in order to create a bite for the mortar of the ceiling.

The author is rather vague on the direct relationships – linear or otherwise? – between the length, height and width, and the resistance of a beam. He clearly didn't follow Galileo's relationship between the resistance and the square of height of the section (which he must have known via Musschenbroek's book), a proposition which was however put into doubt by experimental research (by Buffon, among others) for so-called elastic solids, such as wood [Diderot and d'Alembert 1751, II: 302]. However, due to the extremely vague formulation and the lack of references, it is not possible to determine if the author of the *Tentare licet* paper knew and understood these writings.

The first figure (fig. 1) of *Tentare licet* illustrates clearly that the geometrical side condition proposed by de Nieuport – a beam resting on its two ends – is not well understood. In fact, the author represents in general a condition closer to reality: a beam wedged in the wall at both ends.

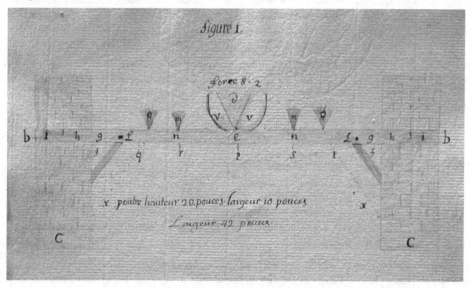

Fig. 1

In the second chapter the author deals with the causes of rupture and the places where they occur. He formulates the following principles: (a) that a force must provoke a sharp angle in the longitudinal fibers, and as a consequence, provoke decomposition of the elementary molecules; (b) that a rope cracks in the middle in the case of symmetrical forces and at the end in the case of asymmetrical forces; and (c) that fibers crack in points of excessive force. Fig. 3 of the paper illustrates this chapter and shows an elementary modeling of the problem with ropes and weights.

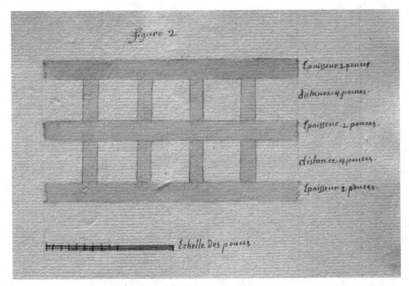

Fig. 2

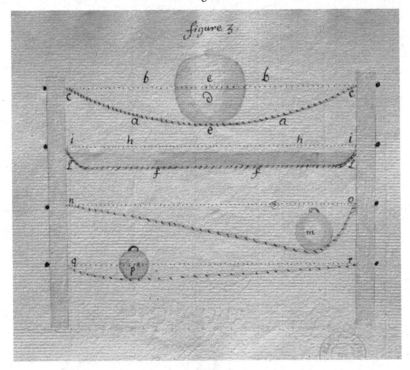

Fig. 3

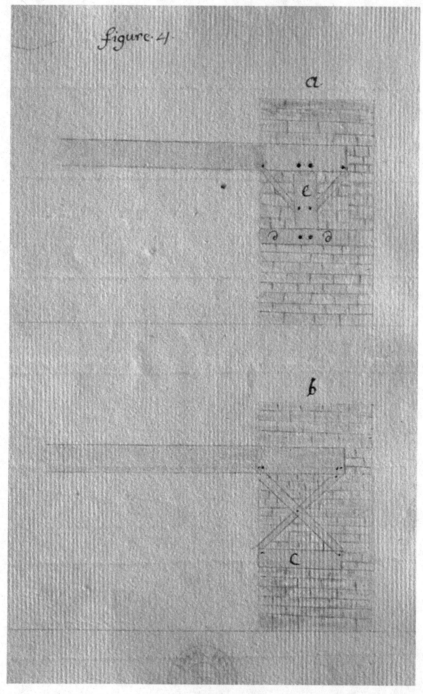

figure. 4

Fig. 4

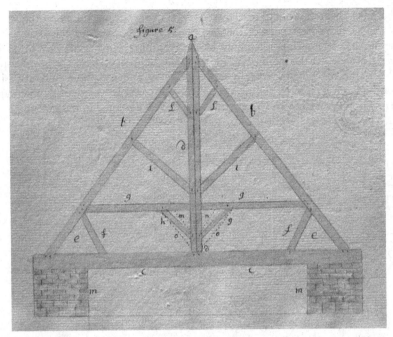

Fig. 5

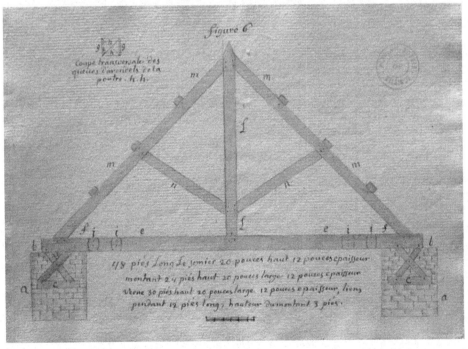

Fig. 6

In the third chapter, the author shows ways of strengthening the beam, based on the analysis of the first chapter: strengthening the supporting ends of the beams (fig. 4); reducing the span by attaching the lowest beam of a timber frame at the ends through *liens pendants* (fig. 5) and at the middle (fig. 6); the same technique is used on the other parts of the timber frame; reducing the span by supplementary wooden supports and increasing the height of the beam section by adding a supplementary piece of wood above the beam (fig. 7). The other figures (figs. 8 and 9) illustrate case studies discussed in the appendix.

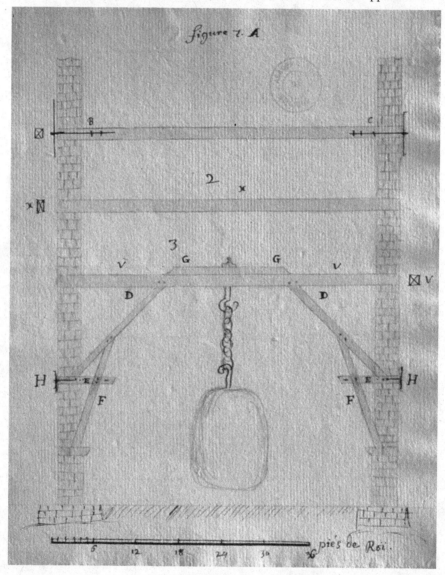

Fig. 7

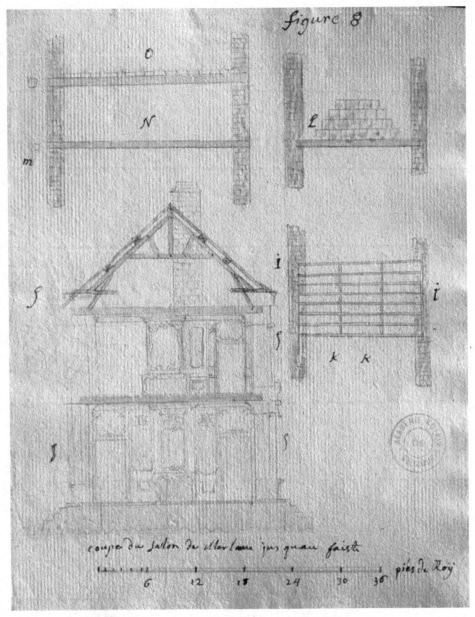

Fig. 8

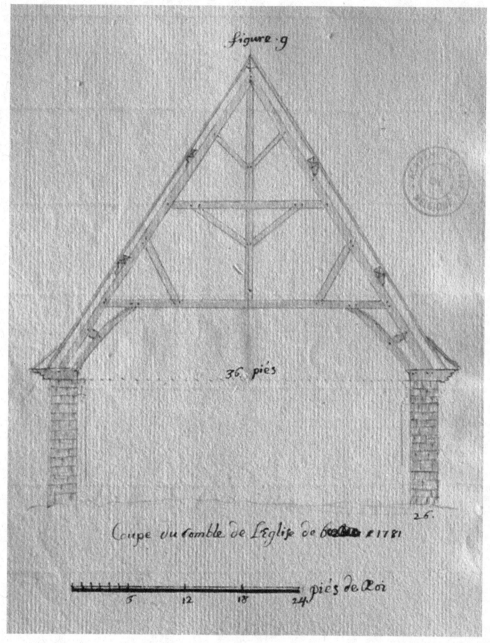

figure .9

36. piés

26.

Coupe du comble de L'Eglise de ~~bordeaux~~ e 1781

5   12   18   24   piés de Roi

Fig. 9

It is stunning that none of these papers, even the more elaborated French one, fully exploited the available literature. That it was difficult to obtain the very specialized publications on the theoretical modeling and mathematical treatment of the problem can perhaps be an excuse, but references to the vast and related field of research, such as that of the experiments on the strength of wood, which became a popular subject in experimental physics, are lacking as well, as are references to the pioneering work of Galileo, or other works on statics. This is all rather painful, because this knowledge was easily available in the articles "wood" and "beam" of the *Encyclopédie ou dictionnaire raisonné des sciences, des arts et des métiers* [Diderot and d'Alembert 1751; Diderot and d'Alembert 1765]. Even extremely popular reference tables of wood sections (height and width of a beam in relation to the length) – proposed in Bullet's *Architecture pratique* [1691, 222] and in De la Hire's edition of Mathurin Jousse's *Traité de Charpenterie* [1702, 190-191], and quoted over and over again in architectural manuals [Gauthier 1716, 50; Jombert 1728, I, 32-33], as well as in the engineering treatise *par excellence*, Bernard Forest de Bélidor's *La Science des ingénieurs* [1727, II, 30][17] – aren't mentioned. It must indeed be asked if the Academy succeeded in reaching its intended audience with this question, because almost none of the relevant works in the field of architecture, engineering, strength of wood and statics seem to be used.

In fact, the basic challenge of the prize question – to theorize the problem, that is, to make an abstraction of the particular wooden part with its deficiencies – seems to constitute for those authors an almost insurmountable problem. Even in the French *mémoire*, whose author at least used some references, the level of abstraction remains low: the qualitative remarks weren't put into a formula; neither were the geometrical aspects of the problem exploited, nor was the available data on the strength of wood taken into consideration. As a result, even the most elaborated paper ended up as no more than a small instruction book for carpentry.

Conclusion

The physics question of the Brussels academy of 1783 on the theory of the beam, proposed by Viscount de Nieuport illustrates marvelously a historical moment in the local history of building science: a new schism presents itself, a schism between an new "mathematical" engineering culture (characterized by the modeling and the calculus of structures), represented in the Southern Low Countries by such "isolated incidents" as the Academy's papers on vaults and the formulation of the beam question, and an existing, intuitive, practice-based artisan's culture, which would remain for a long time in a sort of autarchic condition, uninfluenced even by architectural and engineering manuals.

The answers to this prize question make us aware of the novelty of the rising mathematical approach to solving mechanical problems. The numerous hints and suggestions hidden in the precise formulation of the question weren't evident enough for the general public – nor even the "learned" public – to decode. In this case, one can also recognize a serious problem in the transfer of knowledge, as well as the difficulty that the Academy had in reaching its audience (engineers/architects, mathematicians, and physicists). It would be decades before the new mathematical approach, which was intended to make structures calculable through "modeling", the use of advanced mathematics, and tested material characteristics would have a concrete effect in Belgian building construction.

Acknowledgment

Figs. 1-9 are reproduced with the kind permission from Bruxelles, Académie des Sciences, des Lettres et des Beaux-Arts de Belgique, Archives Anciennes 214.

Notes

1. Bruxelles, Académie Royale des Sciences, des Lettres et des Beaux-Arts de Belgique, Archives Anciennes (further abbreviated as ARB, AA), 213 "N. 1 *De oordeelen der menschen worden heel verschillig gevonden, op eene merkweerdige zaeke*"; ARB, AA 214 "N. 2 *Tentare licet*"; ARB, AA 215 "N.3 *Alle deeze t'saemen-gevoegde/ is hart met hart*"; ARB, AA 216 "N.4. *Den Zang verheugd/ In eer en deugt/ Hunne Zoetigheijt/ Maeckt ons verbleijt*".

2. For the short overview with bibliography, we refer to our institutional approach of the subject [Van de Vijver 2003b].

3. On the Theresian Academy, especially its *classe physique* see: [Académie royale de Belgique 1872]; [Mailly 1883]; [Académie royale de Belgique 1922]; [Marx 1977]; [Maton, Bockstaele, Van Hoof, Van Dormael, Vandepitte and Gullentops 1983]; [Roegiers 1983].

4. *Le premier objet sans doute que notre Société (littéraire) naissance doit envisager est de se rendre utile à la patrie* [Mailly 1883, II, 1]. All the translations in this article are by the author.

5. On de Nieuport see: [de Gavre 1827, vii]; [Garnier n.d.]; [de Rieffenberg 1829, IV: 142 (notice nécrologique)]; [Bigwood 1889]; [Bockstaele 1983: 106-108]. On de Nieuport's calculus of vaults, see: [Radelet-de Grave 1995]; [Van de Vijver 2003a: cat. n. 36-37 ("De Nieuport en het berekenen van gewelven")]; [Mailly 1883, II, 110 (session of 12 October 1780)]..

6. *Ce grand mémoire me paraît, à tous égard, bien fait, curieux et intéressant, et je n'hésite pas de dire qu'il mérite d'avoir place dans le second volume de nos Mémoires, par préférence aux deux autres du même auteur, qui traitent de sujets moins utiles que celui-ci* [Mailly 1883, II, 86-87 (session of 18 May 1778)].

7. [Radelet-de Grave 1995: 157-161 ("V. 1778. Charles François le Prud'homme d'Hailly, Vicomte de Nieuport, Essai analytique sur la méchanique des voûtes présenté le 18 mai 1778")].

8. [Mailly 1883, II, 136-137 (Session of 18 March 1784); 309-310 (question in French); 314]; [Van de Vijver 2003a: cat. n. 38 ("De prijsvraag van 1784 over het berekenen van een vrij opgelegde balk op twee steunpunten")].

9. *Développer la théorie des poutres qui reposent par leurs extrémités sur deux points d'appui, en les considérant dans l'hypothèse la plus conforme à la nature, c'est-à-dire, comme des amas de fibres pesantes, extensibles, élastiques et unies entre elles dans toute leur longueur. Déduire de cette consideration la cause de leur rupture et l'endroit où elle doit se faire dans les différentes cas, par rapport aux différentes situations des masses dont ces poutres pourraient être chargées, et déterminer en conséquence le meilleur emploi des liens pendants* [Mailly 1883, II, 309]. The contemporary Dutch translation of the question is: *Te ontknoopen de Theorie der Balken, die door hunne beyden eynder op twee steunsels rusten, de zelve aenmerkende in de Hypothése, meest overeenkomstig aen de nature, dat is te zeggen: als versaemeling van wegende vesels, extensibel, elastig en onder elkanderen vereenigd in hunne geheele lengde. Van deze aenmerkinge voord te bringen de oorsaeke van hunne breuke, en de plaetse, alwaer de selve in de verscheyde gevallen moet geschieden ten opsigte van de verscheyde gelegentheden der gewichten, waermede die balken sauden mogen gelaeden zyn, en gevolgendlyk te bepaelen het beste gebruyk van de hangende banden* [*Den Vlaemschen indicateur ofte aenwyzer der wetenschappen en vrye-konsten*, 1783, Ghent, X, 291-292 ("Akademische Berichten"), 292]; also quoted in ARB, AA 213.

10. On this aspect see: [Licoppe 1996, 195-242 ("Dure comme du bois: la preuve utilitaire et la question de la résistance des solides au Siècle des Lumières")].

11. *A ne point distribuer le prix, à ne faire aucune mention honorable et à abandoner la question* [Mailly 1883, II, 310].

12. To the author of the paper *Alle deeze t'saemen-gevoegde/ is hart met hart*, a beam on two supports symbolizes the death penalty through hanging (sic). He considers this penalty to be the

only one apt to punish repetitive disloyalty to the Ten Commandments; for a first offence to the Ten Commandments, he judges corporal punishment as being sufficient.

13. *Een "suyveren balk sonder weeren ofte quaede plekken, ende den selven aen merkende als bij het vraegstuk bepaeld"* (ARB AA 213).

14. *... geenen vasten regel op alle balcken, want het verschil van hout, is seer ongelijck van den eenen balck tot den anderen, en daerom onmogelijk daer van eenen vasten regel te geven* (ARB AA 216).

15. *... in de vergaedering en ofte jointuren der maillie/ Naer welke plaetse het gewicht altijdt sijne toevlucht neemt* (ARB AA 213: § 7).

16. [Musschenbroek 1739]; [Nollet 1745-1755]. However, it is important to mention that the Theresian Academy didn't share the general opinion, acquired by Nollet through his written work, that he was a "great physicist" (*grand physicien*), or a respectable man, handy in experiments and physical machines (*cela n'est pas conforme à l'idée générale qui s'est formée de cet homme respectable, de cet habile faiseur d'expériences et de machines physiques, d'après ses propres ouvrages*) [Mailly 1883, II, 136-137 (session of 18 March 1784)]..

17. The table in Briseux [1761, II, 93] is different.

Bibliography

ACADÉMIE ROYALE DE BELGIQUE. 1872. *Centième anniversaire de fondation (1772-1872)*. Brussels: Académie royale de Belgique.

――――. 1922. *L'Académie royale de Belgique depuis sa fondation (1772-1922)*. Brussels: Académie royale de Belgique.

BERNOUILLI, James I. 1705. Véritable hypothèse de la résistance des solides, avec la démonstration de la courbure des corps qui font ressort. *Histoire et Mémoires de l'Académie royale des sciences*, Paris: 176-186.

BIGWOOD, G. 1889. Nieuport (Charles-François-Ferdinand-Florent-Antoine le Prudhomme d'Hailly, Vicomte de), *Biographie nationale*, XV: col. 712-718.

BOCKSTAELE, P. 1983. De Wiskunde. Pp. 105-110 in MATON, J., BOCKSTAELE, P., VAN HOOF, A., VAN DORMAEL, A., VANDEPITTE, D. and GULLENTOPS, F., De positieve wetenschappen in Vlaanderen 1769-1938, in VERBEKE, G., ed., *De weg naar eigen academiën 1772-1983, colloquium, Brussel, 18-20 november 1982*, Brussels: 105-133, 105-110.

BRISEUX, C.E. 1761. *L'art de bâtir des maisons de campagne, où l'on traite de leur distribution, de leur constuction, & de leur decoration*. Paris: J.B. Gibert.

BULLET, Jean. 1691. *L'achitecture pratique, qui comprend le detail du Toisé, & du Devis des Ouvrages de Massonnerie, Charpenterie, Menuiserie, Serrurerie, Plomberie, Vitrerie, Ardoise, Tuille, Pavé de Grais & Impression*. Paris: Estienne Michallet.

DE GAVRE. 1827. Eloge du commandeur de Nieuport, prononcé par le prince de Gavre, président de l'Académie dans la séance du 6 octobre 1827, *Nouveaux mémoires de l'Académie royale des Sciences et Belles-Lettres de Bruxelles*, IV : i-xii.

DE LA HIRE, Philippe, ed. 1702. *L'art de charpenterie de Mathurin Jousse. Corrigé & augmenté de ce qu'il y a de plus curieux dans cet Art, des Machines les plus necessaries à un Charpentier. Par Mr. D.L.H. Le tous enrichy de Figures & de Tailles douces*. Paris: Tomas Moette.

DE RIEFFENBERG, Baron F. 1829. *Archives pour servir à l'histoire civile et littéraire des Pays-Bas, faisant suite aux archives philologiques*, IV, Brussels.

DIDEROT, Denis and Jean D'ALEMBERT, Jean, eds. 1751. *Encyclopédie ou dictionnaire raisonné des sciences, des arts et des métiers, par une société de gens de lettres*. Paris, II: 297-307 (article "Bois").

――――. 1765. *Encyclopédie ou dictionnaire raisonné des sciences, des arts et des métiers, par une société de gens de lettres*, ... Paris, XIII: 254 (article "Poutre").

FOREST DE BÉLIDOR, Bernard. 1727. *La science des ingénieurs dans la conduite des travaux de fortification et d'architecture civile*. Paris: Claude Jombert.

GARNIER, J.-G. n.d. Sur le mémoire de M. le commandeur de Nieuport ayant pour titre : des notions fondamentales en géométrie tant élémentaire que transcendante, *Annales belgiques des sciences, arts et littératures*, IV: 378-383.

GAUTHIER, Henri. 1716. *Traité des ponts, ou il est parlé de ceux des romains & de ceux des Modernes; de leurs manières; tant de ceux de Maçonnerie, que de Charpente; & de leur disposition dans toute sorte de lieux.* Paris: André Cailleau.

JOMBERT, Claude. 1728. *Architecture moderne ou l'art de bien bâtir pour toutes sortes de personnes tant pour les maisons des particuliers que pour les palais*, Paris: Claude Jombert.

DE NIEUPORT, Charles-François-Ferdinand le Prud'homme d'Hailly, Viscount. 1780. Essay analytique sur la méchanique des voûtes présenté le 18 mai 1778, *Mémoires de l'Académie impériale et royale des sciences et des belles-lettres de Bruxelles* II: 41-137.

————. 1783. Mémoire sur la propriété prétendue des voûtes en chaînettes, presenté le 6 novembre 1780, *Mémoires de l'Académie impériale et royale des sciences et des belles-lettres de Bruxelles*, IV: 1-26.

LICOPPE, Christian. 1996. *La formation de la pratique scientifique: le discours de l'expérience en France et en Angleterre (1630-1820)*, (Textes à l'appui: série antropologie des sciences et des techniques, CALLON, M. and LATOUR, B., eds.), Paris: éditions la découverte.

MAILLY, E. 1833. *Histoire de l'académie impériale et royale des sciences et belles-lettres de Belgique*, (Mémoires couronnés et autres mémoires publiés par l'Académie royale des sciences, des lettres et des beaux-arts de Belgique, collection in-8°, XXXIV en XXXV), Brussels.

MARX, J. 1977. L'activité scientifique de l'Académie impériale et royale des Sciences et Belles-Lettres de Bruxelles 1772-1794, *Etudes sur le XVIII* siècle, IV: 49-61.

MATON, J., BOCKSTAELE, P., VAN HOOF, A., VAN DORMAEL, A., VANDEPITTE, D. and GULLENTOPS, F., 1983. De positieve wetenschappen in Vlaanderen 1769-1938, in VERBEKE, G., ed., *De weg naar eigen academiën 1772-1983, colloquium, Brussel, 18-20 november 1982*, Brussels: 105-133.

MUSSCHENBROEK, Pieter van. 1739. *Essai de physique. Avec une description de nouvelles sortes de Machines Pneumatiques, et un recueil d'Experiences.* Leiden: Samuel Luchtmans, 1739 (French edition of *Institutiones physicae conscripta in usus academicos*, 1734. Leyde: S. Luchtmans).

NOLLET, Jean-Antoine, 1745-1755. *Leçons de physique expérientale*, Paris: frères Guérin.

OZANAM, Jacques. 1778. *Récréations mathématiques et physiques, qui contiennent les problèmes & les questions les plus remarquables, & les plus propres à piquer la curiosité, tant des mathématiques que de la physique.* Paris: Cl. Ant. Jombert, fils aîné.

RADELET-DE GRAVE, Patricia. 1995. Le 'de curvatura fornicis' de Jacob Bernoulli ou l'introduction des infiniments petits dans le calcul des voûtes. Pp. 141-163 in *Entre mécanique et architecture - Between mechanics and architecture*, P. Radelet-de Grave and E. Benvenuto, eds. Basel-Boston-Berlin: Birkhäuser.

ROEGIERS, Jan. 1983, "De academie van Maria-Theresia in historisch perspectief". Pp. 29-42 in *De weg naar eigen academiën 1772-1983, Colloquium, Brussel, 18-20 november 1982*, G. Verbeke, ed. Brussels.

VAN DE VIJVER, Dirk. 2003a. *Ingenieurs en architecten op de drempel van een nieuwe tijd, 1750-1830*, Leuven: Leuvense Universitaire Pers.

————. 2003b. From Nieuport to Mangel: An institutional history of building science in Belgium, 1780-1930. Pp. 2055-2063 in vol. 3 of *Proceedings of the First International Congress on Construction History Madrid, 20th-24th January 2003*, Santiago Huerta, ed., Madrid: Instituto Juan de Herrera, Escuela Técnica Superior de Arquitectura.

About the author

Dirk Van de Vijver is an engineer / architect (Katholieke Universiteit Leuven, 1992). He obtained a D.E.A. en Histoire de l'Art et d'Archéologie at the Université de Panthéon-Sorbonne in Paris (1993), and a Ph.D. in Leuven on *Les relations franco-belges dans l'architecture des Pays-Bas méridionaux, 1750-1830* (2000). He worked on different research projects on the history of eighteenth- and nineteenth-century architecture, construction, building professions, building administrations, building organization, and building sciences in Belgium in an international perspective. He has published, among other things, a book on the engineering and architectural profession in Belgium, *Ingenieurs en architecten op de drempel van een nieuwe tijd, 1750-1830* (Leuven: Universitaire Pers Leuven, 2003) and an article on "Construction History in Belgium" (in: Antonio Becchi, Massimo Corradi, Federico Foce & Orietta Pedemonte, ed., *Construction History. Research Perspectives in Europe,* Kim Williams Books, 2004). From 1993 through 2006 he worked as a researcher at the Architectural History and Conservation research group of Leuven University. As of July 2006, he is a post-doctoral researcher at Utrecht University (Onderzoeksinstituut Geschiedenis en Cultuur), where he studies the architectural relationships between the Netherlands and the Balticum, especially in Gdansk, around 1600, within the context of an international research project "The Low Countries at the Crossroads. Netherlandish Architecture as an Export Product in Early Modern Europe (1480-1680)".

Olivier Baverel

Ecole Nationale des Ponts et Chaussées
LAMI
6 et 8 avenue Blaise Pascal
Cité Descartes - Champs sur Marne
77455 Marne La Vallee Cedex 2
FRANCE
baverel@lami.enpc.fr

Hoshyar Nooshin

School of Engineering (C5)
University of Surrey
Guildford, Surrey, GU2 7XH UK
H.Nooshin@surrey.ac.uk

Research

Nexorades Based on Regular Polyhedra

Abstract. The objective of this paper is to discuss the characteristics of nexorades based on regular polyhedra. An important application of nexorades is for shelters of various sizes and shapes for temporary or permanent purposes. In such a shelter, the structural skeleton is provided by a nexorade and the cover is provided by a membrane material.

Keywords: mutually supporting elements, form finding, polyhedron, nexorades

1. What is a nexorade?

Consider the structure shown in fig. 1. This structure is made from wooden elements using an 'interwoven pattern' as shown. The wooden elements are about one metre long and have a 27mm×27mm square cross-section. The connections are made by passing pieces of wires through holes in the elements.

This structure was designed and constructed by Olivier Baverel and exhibited at the University of Nottingham, UK, during the third Colloquium of the IASS Working Group on Structural Morphology in August 1997.

Fig. 1. A nexorade made from wooden elements, exhibited in August 1997 at the University of Nottingham UK

The structure shown in fig. 1 is an example of a nexorade [Baverel, et. al 2000] or a 'multi-reciprocal grid' [Baverel and Saidani 1998]. Each one of the elements that constitute a nexorade is referred to as a 'nexor'. A nexor has four connection points, two of which are at the ends of the nexor and the other two are at two intermediate points along the nexor. The term 'nexor' is a Latin based word meaning a 'link' and the term 'nexorade' implies an 'assembly of nexors'.

1590-5896/07/020281-18 DOI 10.1007/s00004-006-0043-0

2. Engagement windows and fans

The arrangement of nexors in a nexorade gives rise to a number of 'openings' that are referred to as 'engagement windows'. A typical engagement window in the nexorade of fig. 1 is of the form shown in fig. 2. The assembly of the nexors that forms an engagement window is referred to as a 'fan' or a 'reciprocal frame' [Popovic 1996].

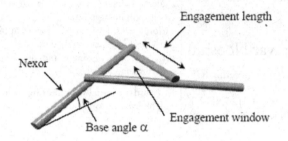

Fig. 2. A fan with a triangular engagement window

The length of each side of the engagement window is referred to as the 'engagement length'. Also, the ratio of the length and the total length of a nexor is referred to as the 'engagement ratio'. Typical values for the engagement ratio are normally in the range 0.1 to 0.4.

Nexors are usually tubular and are connected together using some sort of swivel connector. A view of a fan with scaffolding tubes as nexors and swivel couplers as connectors is shown in fig. 3.

Fig. 3. A fan with scaffolding tubes as nexors and swivel couplers as connectors

Another example of a nexorade is shown in fig. 4. Each of the fans in this nexorade involves four nexors. A typical fan of the nexorade of fig. 4 is shown in fig. 5. The number of nexors constituting a fan is referred to as the 'valency' of the fan. Thus, the fan of fig. 2 has valency 3 and the fan of fig. 5 has valency 4.

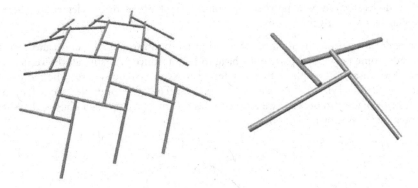

Fig. 4. A nexorade with 4-nexor fans   Fig. 5. 4-nexor fan

It is not necessary for all the fans in a nexorade to have the same valency. For example, the nexorade shown in fig. 6 has two fans with valency 3 and two fans with valency 4.

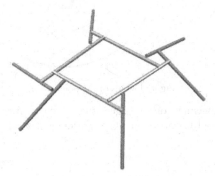

Fig. 6. A nexorade involving fans with different valencies

3. Method for creating nexorades based on reglar polyhedra

A method referred to as the method of rotation has been used to generate nexorades based on regular polyhedra. Consider the dodecahedron shown in fig. 7.

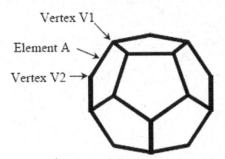

Fig. 7. Dodecahedron

The dodecahedron is a polyhedron composed of 30 identical elements, where each vertex such as V1 joins three elements.

In order to explain the method of rotation for transforming the dodecahedron into a nexorade, some polyhedral particulars have to be introduced. Firstly, all the vertices of the dodecahedron are on a sphere that is referred to as the circumsphere. The radius of this sphere is denoted by Rc as shown in fig. 8. The midpoints of the edges of the dodecahedron are also on a sphere referred to as the intersphere. The radius of this sphere is denoted by Ri, as shown in fig. 8.

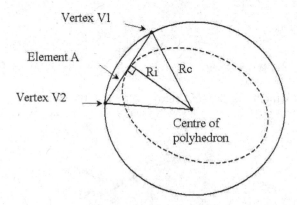

Fig. 8. Details of the polyhedron

Fig. 9 shows an arrangement of the edges of the dodecahedron joining at vertex V1. All the three edges A, B, and C shown in fig. 9 are tangent to the intersphere at their mid point.

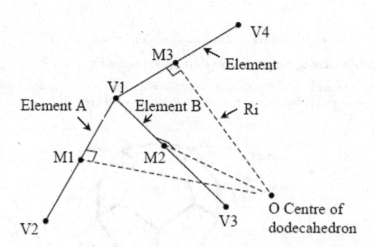

Fig. 9. Details of vertex V1

Fig. 10 shows a plane P1 that contains element A and is tangent to the intersphere. The direction of the normal vector of plane P1 at point M1 will pass through the centre of the dodecahedron which is taken as the origin of the coordinate system.

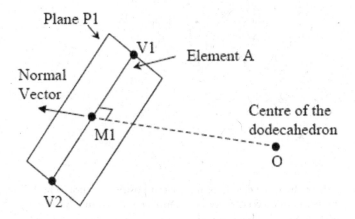

Fig. 10. Detail of element A

To use the method of rotation, element A is rotated by an angle θ around point M1 in plane P1, as shown in fig. 11. The same procedure is repeated for elements B and C that are connected to V1, as shown in fig. 12.

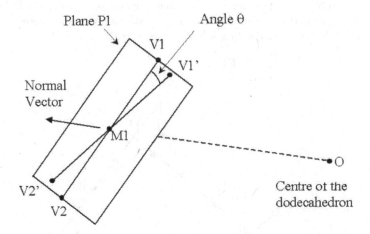

Fig. 11. Rotation of element A in plane P1

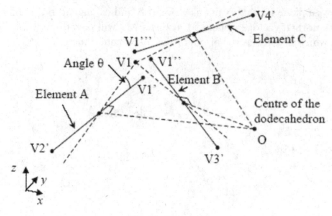

Fig. 12. Vertex V1

The V1 ends of the three elements are now in positions denoted by V1', V1" and V1"'. In order to represent the eccentricity between elements A and B, a line perpendicular to both lines has to be found. This line is the common perpendicular which is also the shortest distance between the two lines. The method to find the common perpendicular is explained in the next section.

Calculation of the eccentricity. The lines denoted as 'line A' and 'line B' in fig. 13 are meant to indicate the directions of elements A and B in fig. 12, respectively. Consider the points Q1 and Q2 lying, respectively, on the lines A and B as shown in fig. 13.

The line Q1Q2 is meant to represent the eccentricity between line A and B. Line Q1Q2 has to be perpendicular to both lines A and B, thus the vector product between the vector V1'V2' and V1"V3' will give the direction of the line Q1Q2.

$$\vec{N} = \vec{V1'V2'} \times \vec{V1''V3'} \tag{1}$$

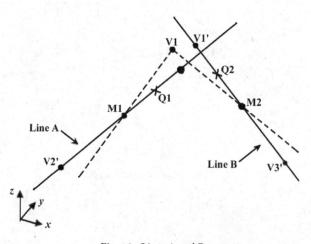

Fig. 13. Lines A and B

Now, consider the plane Q1Q2V1′V2′, which is denoted by Π1, fig. 14. Π1 contains V1′V2′ and Q1Q2, which is parallel to the vector N. Point Q2 is the intersection of plane Π1 and Line B, as shown in fig. 14.

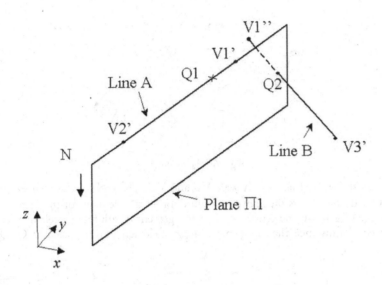

Figure 14. Location of point Q2

Consider now the plane Q1Q2V1″V3′, which is denoted by Π2. Π2 contains V1″V3′ and Q1Q2, which is parallel to the vector N. Point Q1 is the intersection of plane Π2 and line A, as shown in fig. 15.

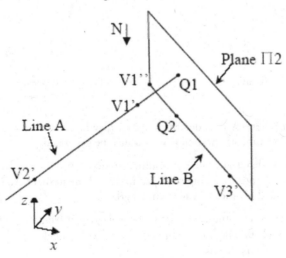

Fig. 15. Location of point Q1

The eccentricity is then equal to the distance between points Q1 and Q2, as shown in fig. 16.

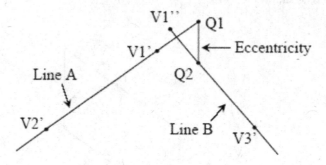

Fig. 16. The eccentricity

Originally the ends of element A were V1' and V2'. The ends of this element are now V2' and Q1. If one repeats the above procedure for all the elements of the vertex, the fan shown in fig. 17 will then be generated. The engagement length for each element can now be calculated. For instance, the engagement length of element A is the distance Q1Q6.

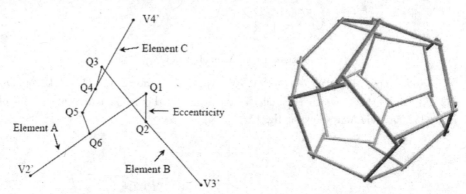

Fig. 17. Resulting fan   Fig. 18. Rendered dodecahedric nexorade

The procedure for vertex V1, as described above, may be repeated for all the vertices of the polyhedron, a 'dodecahedric nexorade' would then be generated.

Fig. 18 shows a rendered view of the resulting dodecahedric nexorade. The particular nexorade shown corresponds to an angle θ equal to 5° and the nexors have identical length, identical eccentricity and identical engagement length.

Fig 19 shows the variations of the eccentricity, the engagement length and the nexor length with respect to the angle θ. The horizontal axis represents the angle θ and the vertical axis represents the value of the parameters.

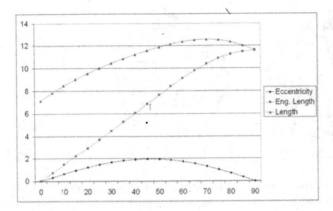

Fig. 19. Variations of the eccentricity, engagement length and nexor length with respect to the angle θ for a dodecahedron with a circumradius of 10 units

Figs. 20 – 28 show nexorades with an angle θ equal to 5°, 10°, 20°, 30°, 45°, 60°, 70°, 80° and 85° degrees, respectively.

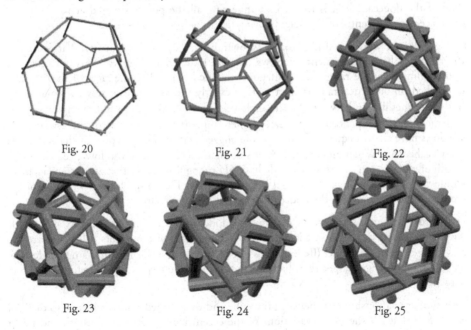

Fig. 20   Fig. 21   Fig. 22

Fig. 23   Fig. 24   Fig. 25

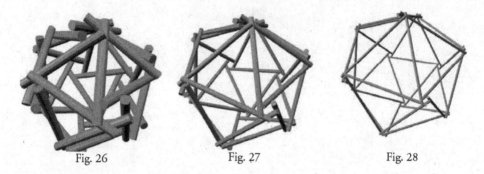

Fig. 26   Fig. 27   Fig. 28

The nexorade shown in fig. 20 is referred to as a 'dodecahedric nexorade' with angle θ equal to 5°. The nexorade shown in fig. 28 is also referred to as a 'dodecahedric nexorade' with angle θ equal to 85°, but this nexorade resembles more an 'icosahedric nexorade' than a 'dodecahedric nexorade'. This phenomenon is a consequence of the fact that the dodecahedron is the dual of the icosahedron. To elaborate, the dodecahedron and the icosahedron have 30 elements. The dodecahedron has 20 vertices whereas the icosahedron has 20 faces. The dodecahedron has 12 faces whereas the icosahedron has 12 vertices. The dual of the dodecahedron is found by connecting all the points situated at the centre of each face with the centre of the neighbouring faces.

Consider the 'dodecahedric nexorade' shown in fig. 20. The 3-nexor fans that constitute the nexorade have engagement windows that are rather like triangles. This nexorade has also openings that are rather like pentagons, these correspond to the faces of the 'mother' dodecahedron. The nexorade is then neither exactly a dodecahedron nor an icosahedron but it is something in between. When angle θ is increased the size of the triangles increase and the size of the pentagons decrease. When angle θ is equal to zero, the engagement windows have an area equal to zero, thus the engagement windows represent the vertices of a dodecahedron. When angle θ is 90°, a side of the engagement window is equal to the length of a nexor and thus the nexorade has only triangular 'faces', that are the engagement windows. The nexorade in this case is an icosahedron. This example shows the use of the concept of duality between the dodecahedron and the icosahedron.

In order to name this type of nexorade, a dodecahedron transformed into a nexorade with an engagement length smaller than half of the length of the nexors will be referred to as a 'dodecahedric nexorade' (θ<36.28°), a dodecahedron transformed into a nexorade with an engagement length larger than half of the length of the nexors will be referred to as an 'icosahedric nexorade'.

A similar relationship exits between the cube and the octahedron, which are also duals of each other. Fig. 29 shows the variations of the eccentricity, engagement length and nexor length of a 'cubic nexorade' with respect to the angle θ. The horizontal axis represents the angle θ and the vertical axis represents the value of the other parameters.

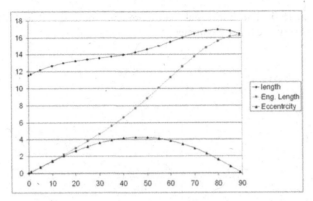

Fig. 29. Variations of the eccentricity, engagement length and nexor length with respect to the angle
θ for a cube with a circumradius length of 10 units

Figs. 30 to 33 show nexorades with an angle θ equal to 10°, 25°, 50° and 80°, respectively. The nexorades with θ equal to 10° and 25° should be referred to as 'cubic nexorades' (θ < 41.93°). The nexorades with angle θ equal to 50° and 80° should be referred to as 'octahedric nexorades', as their engagement lengths are larger than half of their nexors' length.

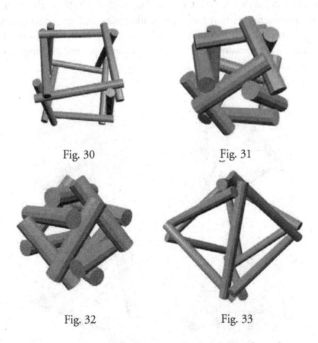

Fig. 30

Fig. 31

Fig. 32

Fig. 33

The remaining regular polyhedron, the tetrahedron, has four faces and four vertices. This means that the tetrahedron will be the dual of itself. Fig. 34 shows the variations of the eccentricity, engagement length and nexor length of a 'tetrahedric nexorade' with respect to the angle θ. The horizontal axis represents the angle θ and the vertical axis represents the value of the other parameters.

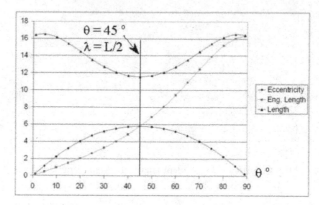

Fig. 34. Variations of the eccentricity, engagement length and nexor length with respect to the angle θ for a tetrahedron with a circumradius length of 10 units

Figs. 35 – 38 show 'tetrahedric nexorades' with an angle θ equal to 10°, 30°, 45° and 80°, respectively. The nexorades with angle θ equal to 10° and 80° are identical except that the style of their fans are different, that is, the one with θ = 10° is the mirror of the one with θ = 80°. The arrangements that look rather like faces and the engagement windows of the nexorade shown in fig. 37 have the particular property of being of identical size. Note also that in this figure some nexors are parallel to one another and the engagement length is equal to half of the nexor length.

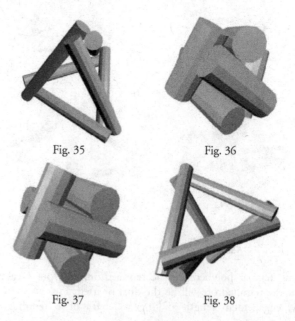

Fig. 35

Fig. 36

Fig. 37

Fig. 38

In conclusion to this study of the method of rotation applied to regular polyhedra, one may state that there exist only three regular polyhedric nexorades. The first of these nexorades is the 'tetrahedric nexorade'. The second nexorade represents a cube or an octahedron depending on whether its angle θ is equal to $0°$ or $90°$. The last nexorade represents a dodecahedron or an icosahedron depending on whether its angle θ is equal to $0°$ or $90°$.

4. Aspect ratio

The ratio between the eccentricity and the engagement length for a regular fan is referred to as the 'aspect ratio' of a fan. This ratio may be equivalently defined as the ratio between the diameter of the cross-section of a nexor in a regular fan and the engagement length of the fan. For a nexor, the term 'aspect ratio' implies the ratio between the diameter of its cross-section and its engagement length.

Consider the elementary configuration (which is a configuration with zero eccentricity and zero engagement length) shown in fig. 39a, where the vertices V2, V3, V4 lie in the xy plane and where elements V2V1, V3V1 and V4V1 have the same length and the same angle with the xy plane. This angle is the base angle α. If one projects the elements V2V1, V3V1 and V4V1 on the xy plane, the angles between the projected elements, denoted by ϕ, will be equal to $120°$, as shown in fig. 39b.

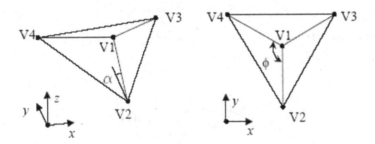

Figure 39 a (above) and b (below) Elementary configuration

If one wants to transform the configuration into a nexorade by translating the elements, one should proceed as follows:

$$\frac{N_z}{\|N\|} \cdot e = \lambda \cdot \frac{V2V1_z}{\left\|\overrightarrow{V2V1}\right\|} \qquad (2)$$

where:

- N is the vector product between V2V1 and V3V1;
- V3V1$_z$ is the z component of vector N;
- e is the eccentricity;
- is the engagement length.

The ratio between the eccentricity and the engagement length is, therefore, given by

$$\frac{e}{\lambda} = \frac{V2V1_z}{\left\| \overrightarrow{V2V1} \right\|} \cdot \frac{\left\| \overrightarrow{N} \right\|}{N_z} \tag{3}$$

Note that the values of the scalars V2V1z and Nz depend on the orientation in space of elements V2V1 and V3V1. These orientations can be expressed in terms of two variables, namely, the base angle α and the angle ϕ. It is assumed that the length of element V2V1 and V3V1 are equal to one unit length. The components of vectors V2V1 and V3V1 are as follows:

$$V\overset{\rho}{2}V1 = \begin{pmatrix} 0 \\ \cos(\alpha) \\ \sin(\alpha) \end{pmatrix} \quad V\overset{\rho}{3}V1 = \begin{pmatrix} -\cos(\alpha)\cdot\sin(\phi) \\ \cos(\alpha)\cdot\cos(\phi) \\ \sin(\alpha) \end{pmatrix}$$

Vector N, that is, the vector product of V2V1 and V3V1 is as follows:

$$\overset{\rho}{N} = \begin{pmatrix} \cos(\alpha)\cdot\sin(\alpha) - \cos(\alpha)\cdot\sin(\alpha)\cdot\cos(\phi) \\ -\cos(\alpha)\cdot\sin(\alpha)\cdot\sin(\phi) \\ \cos^2(\alpha)\cdot\sin(\phi) \end{pmatrix}$$

Now, the values of the vectors are substituted in Equation (3).

The final equation is:

$$\frac{e}{\lambda} = \tan(a)\cdot\tan(b)\cdot\left(\left(\frac{1}{\sin(\phi)} - \frac{1}{\tan(\phi)} \right)^2 + \frac{1}{\sin^2(\alpha)} \right)^{\frac{1}{2}}$$

For the configuration shown in fig. 39, the value of ϕ is 120° and therefore the values of cos(ϕ) and sin(ϕ) are, respectively, equal to √3/2 and -0.5. The values of cos(ϕ) and sin(ϕ) are re-introduced in the general equation

$$\frac{e}{\lambda} = \left(4\tan^2(a) - 3\sin^2(\alpha) \right)^{\frac{1}{2}}$$

Now, suppose that one wants to use the method of rotation for the creation of a regular polyhedric nexorade. For any angle θ , one can calculate the aspect ratio of a typical fan of a nexorade, by dividing the eccentricity by the engagement length. The variations of the aspect ratios for 'tetrahedric', 'cubic' and 'dodecahedric' nexorades are shown in fig. 40, where the horizontal axis represents the angle θ and the vertical axis represents the aspect ratio.

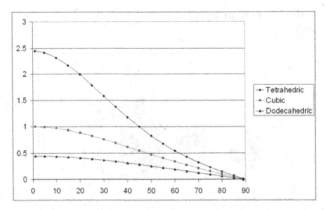

Fig. 40. Variations of the aspect ratios for 'tetrahedric', 'cubic' and 'dodecahedric' nexorades

Consider for instance a fan with an aspect ratio of 0.4. From the curve in fig. 40, it is found that this fan will have a base angle of 19.3°. From the curve in fig. 40, it is found that this fan can be used to create a 'tetrahedric', 'cubic' or 'dodecahedric' nexorade. This means that the nexorades with the aspect ratio of 0.4 for tetrahedric, cubic and dodecahedric nexorades can have fans with identical base angles. The θ angles required to build the fans of 'tetrahedric', 'cubic' and 'dodecahedric' nexorades are 65.5°, 55.5° and 23°, respectively. The only difference between these fans will be the engagement ratio, that is, the ratio between the engagement length and the length. The variations of the engagement ratios for 'tetrahedric', 'cubic' and 'dodecahedric' nexorades are shown in fig. 41, where the vertical axis represents the engagement ratio and the horizontal axis represents the angle θ.

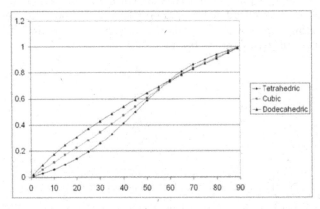

Fig. 41. Variations of the engagement ratios for 'tetrahedric', 'cubic' and 'dodecahedric' nexorades

Consider again, the example where the aspect ratio is 0.4. The θ angles required to build the fans of 'tetrahedric', 'cubic' and 'dodecahedric' nexorades are 65.5°, 55.5° and 23°, respectively. The fans of these nexorades must have engagement ratios of 0.81, 0.67 and 0.34, respectively, as may be obtained from fig. 41.

Fig. 42 shows the fan with an aspect ratio of 0.4 applied to a tetrahedric', 'cubic' and 'dodecahedric' nexorade:

Fig. 42. View of a 'tetrahedric', 'cubic' and 'dodecahedric' nexorades with the same fan

Summary. 'Tetrahedric', 'cubic' and 'dodecahedric' nexorades with aspect ratio of 0.4 have the following particulars:

- Base angle of 19.3° ;

- Engagement ratios of 0.81, 0.67 and 0.34, respectively.

Note that if one considers that the thickness of the elements is equal to zero, a tetrahedron cube and dodecahedron will require a specific value of angle α. In contrast, tetrahedric, cubic and dodecahedric nexorade can be built with a range of value of angle α.

5. History

Nexorades consisting of a single fan have been known for centuries. During the medieval ages, the French architect Villard de Honnecourt [Bowie 1959] provided a solution for the problem of covering a space with beams shorter than the span. He used flat 4-nexor fans with connections of the type shown in fig. 43.

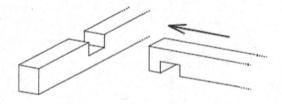

Fig. 43. Connections between nexors in a flat fan

Leonardo da Vinci was also interested in the concept [Leonardo da Vinci 1956]. He made sketches for beam arrangements similar to the 'flat fan' of Villard de Honnecourt. Another architect of the Renaissance period, Sebastiano Serlio, proposed designs, similar to that of Villard de Honnecourt, for spanning spaces with short beams [Serlio 1970]. Less than a century later, John Wallis wrote his *Opera Mathematica* in Latin [1972], where he described a number of flat nexor assemblies.

In recent times, Graham Brown obtained a patent for a roof system consisting of a single fan. Also, D. Gat patented an interesting system consisting of 4-nexor fans [Gat 1978].

Artists are also interested by the concept, recently artist Rinus Roelofs has produced interesting sculptures (see http://www.rinusroelofs.nl).

The nexorades that are discussed in the present paper are the spatial generalisations of the flat beam arrangements of John Wallis. However, the idea of spatial nexorades is not entirely new. For instance, it is known that Leonardo da Vinci had a clear understanding of the workings of the spatial nexorade-like arrangements. This is evidenced by the sketch he produced for a temporary bridge, shown in fig. 44.

Fig. 44. A design for a temporary bridge, suggested by Leonardo da Vinci

Leonardo's bridge design does not quite fit into the family of the nexorades, as described in this paper. However, there are certainly intriguing similarities.

6. Conclusions

This paper has presented a method to generate nexorades based on regular polyhedra. It has been shown that 'tetrahedric', 'cubic' and 'dodecahedric' nexorades can be built with fans identical aspect ratio in other words with the same base angle α.

References

BAVEREL, O., and M. SAIDANI. 1998. The Multi-Reciprocal Grid System. Pp. 66-71 in *Proceedings of the International Conference on Lightweight Structures in Civil Engineering*, Jan B. Obrebski, ed. Warsaw, Poland.
BAVEREL, O., H. NOOSHIN, Y. KUROIWA and G. A. R. PARKE. 2000. Nexodes. *International Journal of Space Structures* 15, 2: 155-159.
BOWIE, T. 1959. *The Sketchbook of Villard de Honnecourt*. Indiana University Press.
GAT, D. 1978. Sigma System. Israeli Patent.
LEONARDO DA VINCI. 1956. *Leonardo da Vinci*. New York: Reynal and Company.
POPOVIC, O. 1996. The Architectural Potential of the Reciprocal Frame. PhD Thesis, University of Nottingham, UK.
SERLIO, S. 1970. *First Book of Architecture by Sebastiano Serlio* (1619). New York: Benjamin Bloom Publishers.
WALLIS, J. 1972. *Opera* Mathematica (1695). New York: Verlag Hildesheim.

Frans A. Cerulus

Instituut voor Theoretische Fysica
Katholieke Universiteit Leuven
Celestijnenlaan 200D
B-3001 Heverlee
frans.cerulus@fys.kuleuven.be

Keywords: pyramids, obelisks,
acoustics, whispering gallery,
golden section

Research

A Pyramid Inspired by Mathematics

Abstract. An eighteenth-century pyramid near Brussels contains intriguing mathematical ratios that suggest they were influenced by Egyptomania of the period. Jesuit priest Athanasius Kircher published several books on mysticism and symbolism that were typical of the times. He also wrote a book on acoustics and described the "whispering gallery", an effect which can be observed inside the Wespelaar pyramid.

Introduction

In the village of Wespelaar, 25 km northeast of Brussels (Belgium) stands a curious little pyramid, in the midst of a luscious private park (fig. 1). It is a *folie* or *fabrique* : a little monument that, in an eighteenth-century landscaped park, holds the attention of the strolling visitor and brings him to a certain mood. The park contains several of these: a temple of Flora, an artificial lake with cave, an *Elyseum* around an obelisk.

Fig. 1. The pyramid in Wespelaar, Belgium

The park and the pyramid date from 1797. They belonged to the Louvain brewer Leonardus Artois, who had acquired a mansion and land in Wespelaar and asked the Brussels architect Ghislain Joseph Henry to design a park of 120 hectares and its *fabriques*.

It is said that the pyramid is a Masonic symbol, but no original plans or documents are extant [Duquenne 2001].

In order to understand the symbolism of the monument we have to keep in mind the general Egyptomania of the period. We shall digress a little on the Jesuit Athanasius Kircher, a noted Egyptomaniac, and then go over to a detailed description and venture some hypotheses on the symbolism of the mathematical ratios hidden in the pyramid.

Egyptomania

Plato, in the dialogue *Timaeus*, attributed to the Egyptian priests the recording of all important events of the past [Plato 1966, §§ 22,23]. Herodotos reinforced the myth that the Egyptians had secret knowledge and the Romans readily believed this. In Imperial Rome (second and third century A.D.) temples to Isis and Serapis were built; at the same time the knowledge of hieroglyphs was being lost in Egypt. Many obelisks were brought to Rome to adorn public squares.

In the Baroque period many of these were rediscovered among the antique ruins and newly erected, often as part of new monuments, in the seventeenth-century papal city. One of the most conspicuous stands on the grotto sculpted by Bernini in the Piazza Navona, the *Fontana dei quattro fiumi*. The concept of the monument was a brainchild of that most memorable man, Athanasius Kircher. This Jesuit was a polyglot, a mathematician and an ethnologist who founded a famous museum. He was very interested in the reports on Buddhism and Confucianism sent home by the Jesuit missionaries and he himself studied the Greek and Arabic sources on the ancient Egyptian civilisation. He is of course only one in a line of scholars who tried to unravel the secrets of the hieroglyphs [Lamy and Bruwier 2005], but his delirious interpretations strongly influenced the esoteric movements of the eighteenth century.

To Kircher, the old pagan sages had inherited, in veiled form, bits of the complete knowledge of the world that Adam had; they were, so to speak, parallel prophets [Pastine 1978]. Of these, he thought the Egyptians had the most profound knowledge and their hieroglyphs were a secret sacred language. He conjectured rightly that the Coptic language was in fact the old Egyptian language, but – as all his contemporaries – he considered hieroglyphs as ideograms, an idea that was reinforced by the then recent discovery of the Chinese system of writing. He followed an intuitive and mystical interpretation of the Egyptian signs, which led him completely astray. He published an extensive work of 1500 pages on the subject [Kircher 1653], which was widely diffused.

Pope Innocent X called on him to interpret the hieroglyphs on the obelisk that was discovered in the circus of Maxentius and Kircher proposed an allegory suggested to him by his interpretation. Bernini executed the monument and when it was completed Kircher wrote an essay where he explained the allegory [Kircher 1650; Rivosecchi 1982]. In short, his views were:

- The small pyramid atop the obelisk symbolizes the Divinity: its apex points to God and the triangular sides refer to the Trinity.

- The obelisk proper represents the world of the angels, through which God communicates with the sublunary world. Its four faces refer to the four elements (fire, air, water, earth) from which all matter is made.

- The obelisk stands on a cave, the symbol of the subterranean world, another pet idea of Kircher, which he published in another memoir [Kircher 1665]. From this world issue the great rivers. In the Bernini fountain the waters flow through four openings that indicate the four great rivers of the four continents: Danube, Nile, Ganges and Rio de la Plata.

A hundred years later, in the "Age of Reason", the belief in secret Egyptian knowledge was greater than ever. Books were published (see, for example, [Pernety 1796]) that were inspired by the Kircher opus. Napoleon Bonaparte believed in the hidden knowledge of the Egyptians and took over a hundred scientists with him on his Egyptian campaign. It was a period where secret societies sprang up throughout Europe and the Masonic lodges borrowed some of their symbols from the Egyptian myths, as well as from the temple of Solomon. To the French Masons the supposed existence of secret knowledge antedating the Bible was an argument for challenging the spiritual dominance of the clerics.

The pyramid : form and orientation

The pyramid is about 7 m high and built from regular rectangular blocks of grey sandstone. The inner chambers and their vaults are in red brick; those were originally plastered, but now the brick is mostly apparent.

There are two striking similarities with the Kircher theories regarding the shape and the position of the building:

- The pyramid is truncated and topped by a little obtuse pyramid, like an obelisk.

- It stands on a cave: the cellar where the ice from the lake was stored.

In that spirit we could look at the pyramid as the symbol of a link between the lower and a higher world

A vertical section through a diagonal is an equilateral triangle (disregarding the truncation). One recalls here that the superimpostion of the equilateral triangle and its reverse form the hexagram, also called Solomon's Seal. To Kircher, probably taking over an older tradition, an equilateral triangle in a circle was the symbol of the "immobile mover", an aspect of the divinity, reproduced in fig. 2.

Fig. 2. The hierarchy of the muses as an alias for the hierarchy of the angels, according to Kircher [1653]

The four corners of the pyramid point approximately to the cardinal points, but not completely: the NS diagonal makes in fact an angle of about 10° to the east with due north (fig. 3).

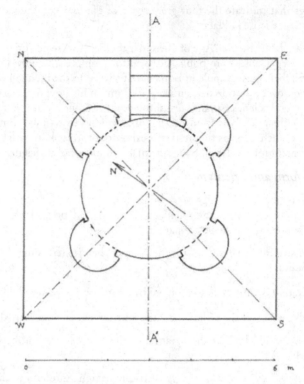

Fig. 3. Base of the pyramid with the projection of the floor of the lower chamber

We offer a tentative explanation. In the pre-Christian era the summer solstice was celebrated, a feast the Church recuperated as St John's day (24 June); in rural areas this is still the occasion to light the St John's fires. In the Masonic tradition, the Orient, the rising sun and even St John play a role. We surmise consequently that the pyramid was in fact oriented with its entrance towards the rising sun on the summer solstice (21 June). At the latitude of Wespelaar (50.53° N) the direction of the rising sun makes an angle of 51.23° with the north, meaning that the NS diagonal should lie 6.23° east of north. This leaves a discrepancy of some 4° to explain. An architect setting out the diagonals on the ground will do so with respect to the north as determined by a compass. He would be aware that there is a local magnetic declination and corrects for it. The most readily available reference on scientific matters at the end of the eighteenth century would have been the *Encyclopédie* of Diderot and d'Alembert, under the voice *aiguille aimantée*. One finds there a table of the declination in Paris for the years 1700 to 1750 when the declination increased linearly from 8°18′ W to 17°15′ W. Extrapolating linearly to 1797 yields an estimated 26° W. The declination, however, was heading for a maximum of 22.5°, which it reached in 1814. Around 1800 it must have been 21.5° with an error of 0.5°. If this hypothesis is true the architect overcorrected the westward declination by some 4.5° to the east, which explains the actual orientation of the pyramid.

The dimensions of the pyramid

The side of the pyramid, near the ground, is 6.10 m. But the actual base is probably larger inasmuch as the level of the soil must have risen through accumulating humus over the last two hundred years, in spite of the excellent maintenance of the park during this time.

At the time of building, the unit of length in Wespelaar was the Brussels foot [Vandewalle 1984]:

$$1 \text{ foot} = 0.27575 \text{ m}$$

Although Belgium had become part of the French republic in 1794, the metric system – barely two years old – was not yet in use.

It is therefore to be expected that the basic measures of the pyramid will be simple numbers when expressed in feet. Everything fits nicely if we assume that the diagonal of the base (which is the side of the equilateral triangle that governs the shape of the pyramid) is 32 feet, i.e. 2^5. For the side of the base this makes

$$32/\sqrt{2} \text{ feet} = 6.24 \text{ m.}$$

The inner chambers. Inside we find a round chamber with slightly conical walls and four niches, which originally held Egyptian vases at the origin. Three steps lead into it, through a small entrance; its floor is 0.74 m above the base (it has recently been re-laid) and has a diameter of 3.35 m. With the original plastering on this was probably 3.31 m, i.e., 12 ft (fig. 4).

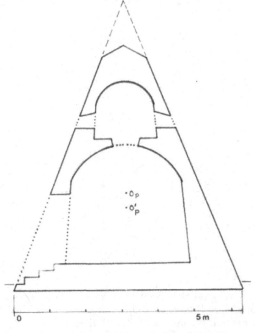

Fig. 4. Vertical section along the line AA' in fig. 3

Fitting the measured heights of the domed ceiling to a simple curve we find it is spherical (or, less probable, parabolic) and certainly not elliptical nor an inverted catenary; the radius of the dome is 6 ft and its centre lies 2.23 m above the base, i.e., practically 8 ft. In the centre of the dome is an oculus with a lower diameter of 0.67m that gives access to a second chamber, which has a floor with a diameter of 1.70 m and a height of 1.50 m. Because the walls are very rough these measures are approximate. This makes the dimensions of the upper chamber practically one half of those of the lower chamber, with a similar conical shape, topped by a spherical cap. There are no niches nor an oculus in the upper chamber, but there are four semicircular openings in the faces of the pyramid; the radius of the semicircles, measured at the outer surface of the building, is 0.28 m, i.e., 1 ft.

Intriguing ratios. Drawing a vertical section of the pyramid, through its apex and parallel to a side (see figs. 5 and 6), we find that:

1. The pyramid is truncated at the upper plane of the cube erected on its base;

2. The distance of the apex of the actual pyramid to the oculus is equal to the distance of the oculus to the floor of the lower chamber;

3. The position of the oculus is such that its distance to the base relates to the side of the base as 1 to 1.618, i.e. the so-called golden ratio: $1 + \sqrt{5} \div 2 = 1.618K \equiv \phi$.

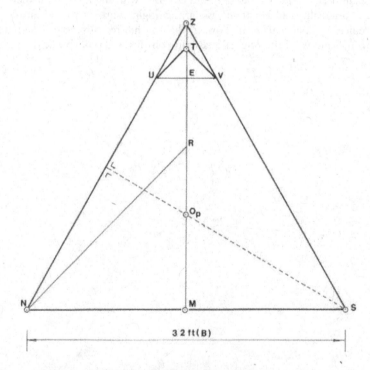

Fig. 5. Design drawing no. 1: section of pyramid through a diagonal, giving an equilateral triangle of 32 Brussels feet; Op is the centre of the circumscribed circle. The faces of the top little pyramid are equilateral triangles.

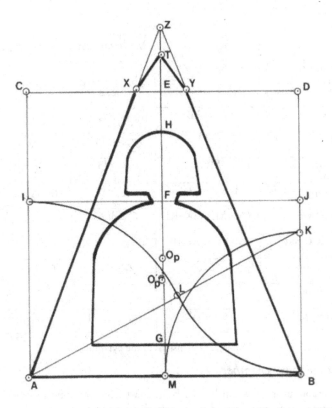

Fig. 6. Design drawing no. 2: vertical section orthogonal to two faces. ABDC is a square. ABJI is a harmonic rectangle. O'p is the centre of the circumscribed sphere to the actual pyramid; this coincides practically with the centre op the spherical cap of the dome

The last characteristic is not surprising. The golden ratio is basic for constructing the pentagram, which we find on the wall of every Masonic lodge or in sorcery manuals as component of a magic circle [Goethe 1808]. It has inspired countless esoteric fantasies since the Renaissance.

The symbolism of the second ratio is less obvious, but let us try one. Let the two chambers represent the onset of the series

$$1 + \tfrac{1}{2} + \tfrac{1}{4} + K = 2$$

Ascending then in spirit by successive steps, in successive chambers, one approaches the ultimate point, which Kircher called God.

We list in Table 1 all the dimensions, in Brussels feet, referring to the segments as defined in figs. 5 and 6.

Name	Segments	Mathematical	Numerical (feet)
Diagonal, rib	NS=NZ=ZS		32
Corner-centre base	NM=MS=MR	NS/2	16
Side base	NR=AB	NS/$\sqrt{2}$	16$\sqrt{2}$=22.63
Height	ZM	NM$\sqrt{3}$	16$\sqrt{3}$=27.71
Base-centre	MOp	MZ/3	16/$\sqrt{3}$=9.24
Height top pyramid	UE=EV=ET	XY/$\sqrt{2}$	16(1-$\sqrt{(2/3)}$) = 2.94
Half side base	AM=MB=BK=KL	AB/2	8$\sqrt{2}$ =11.31
Harmonic to side	AL=AI=FM=JB	AB ($\sqrt{5}$ – 1)/2	($\sqrt{5}$-1)8.$\sqrt{2}$ =13.98
Height cut-off	ZE	ZM-AB	16($\sqrt{3}$-$\sqrt{2}$) = 5.085
Width cut-off	XY		16($\sqrt{2}$-2/$\sqrt{3}$) = 4.15
Height chamber	GF=FT	ET+AB-AI	11.58

Table 1. Dimensions of the pyramid

Acoustics of the lower chamber

Two remarkable properties can be observed, apart from the long reverberation time of sound, characteristic of almost closed rooms with hard walls.

1. A speaker in one niche is more distinctly heard by a listener in the opposite niche than by one in the centre of the room.

2. Speaking in the centre of the room towards the ceiling produces no echo, while speaking sideways or off-centre induces the usual reverberation.

Both effects can be understood by doing geometrical acoustics, the equivalent for sound of geometrical optics; both are good approximations when the wavelength (of sound or light) is short with respect to the characteristic dimensions of the set-up.

In the present case this dimension is about 3 m and the wavelength of speech is around 1 m (frequencies around 300 Hz), making for a tolerable approximation.

The first effect is a clear case of a "whispering dome": the spherical ceiling gathers the sound energy emitted by the speaker and focuses it in a symmetric point, with respect to the axis. The listener on the opposite wall receives all the energy collected by the dome, whereas a person in the centre will only capture the energy falling directly on his auricle. The effect is of course lessened by the reverberating sound where the energy is spread in all directions (fig. 7).

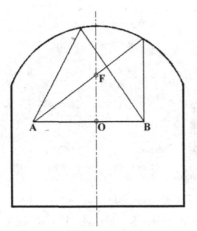

Fig. 7. The rays in a whispering dome. The focus F lies halfway between the centre O
and the mirror

The optical analogy of the second effect would be a concave mirror with a hole in its
centre (the domed ceiling with oculus) facing a plane mirror (the floor). If the rays are
emitted on the axis from a point between the centre of the mirror and the focal point, the
concave mirror focuses them in another point further along the axis; but the plane mirror
reflects them back and there is a region along the axis from which all rays will be reflected
back into the hole. In our case the sound gets into the higher chamber and disappears very
soon into the open through the openings in the wall (fig. 8).

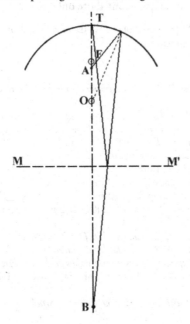

Fig. 8. Sound produced in A is focused in B; the mirror MM' (floor) reflects it in T;
if a hole is present in T the sound disappears without reverberating

Fig. 9. The whispering dome according to Kircher [12]

Were these effects planned? It was Kircher who introduced geometrical acoustics in his *Phonurgia Nova* [Kircher 1673] and described the whispering dome (see fig. 9). It is important to know in this respect that Mr Artois had a close friend and counsellor who lived in the Wespelaar estate and was a former professor of physics at the University of Louvain: Matthieu Verlat. It is almost certain that he discussed the plans with architect Henry. The phenomenon was explicitly mentioned in the course on sound that Verlat taught at the Arts faculty (see [Godaert 1992, 63]).

Professor Verlat was a priest, canon at St Martin's of Liège, and he may have found a religious symbolism in the effects: wise words are understood by a listener in the right disposition while wasted on others and words spoken upwards, in prayer, reach the top, i.e., the Divinity; words spoken at random contribute only noise.

We mention this last hypothesis only as an example how pyramids tend to induce interpretations in Rorschach fashion.

References

DUQUENNE, Xavier. 2001. *Het Park van Wespelaar*. Brussels. (There exists a French version of this work: *Le Parc de Wespelaar*.)

GOETHE, Johann Wolfgang. 1808. *Faust, der Tragödie erster Teil*. Studierzimmer.

GODAERT, Paul. 1992. *Matthieu Verlat, Prêtre-Professeur à «Loven», chanoine ... malheureux, conseiller et AMI de la famille ARTOIS*. Beauvechain.

KIRCHER, Athanasius. 1650. *Obeliscus Pamphilius...* .Rome.

———. 1653. *Oedipus Egyptiacus*. Rome.

———. 1665. *Mundus Subterraneus*. Amsterdam.

———. 1673. *Phonurgia Nova*. Campidonae (Kempten).

LAMY, Florimond and Marie-Cécile BRUWIER. 2005. *L'égyptologie avant Champollion*. Louvain-la-Neuve.

PASTINE, Dino. 1978. *La nascita dell'idolatria. L'Oriente religioso di Athanasius Kircher*. Florence.

PERNETY, Dom Antoine-Joseph. 1786. *Les fables égyptiennes et grecques, dévoilées & réduites au même principe, avec une explication des hiéroglyphes et de la guerre de Troye*. Paris.

PLATO. 1966. *Timaeus* ; trans. R. G. Bury. Cambridge, MA. and London: The Loeb Classical Library.

RIVOSECCHI, Valerio. 1982. *Esotismo in Roma Barocca. Studi sul Padre Kircher*. Rome: Bulzoni editore.

VANDEWALLE, Paul. 1984. *Oude maten, gewichten en muntstelsels in Vlaanderen*, Brabant en Limburg. Gent.

About the author

Frans Cerulus was born in Ghent (Belgium) in 1927. He studied theoretical physics at the universities of Ghent and Basel (Switzerland). He did research in theoretical particle physics in Copenhagen and Geneva (CERN). He became professor of theoretical physics at the University of Louvain (Belgium) in 1964. Since his retirement he has turned to the history of physics. He is co-editor of the collected works of Daniel Bernoulli.

Yannick Joye

Philosophy Department
Faculty of Arts & Philosophy
Ghent University, BELGIUM
yannick.joye@telenet.be

Keywords: Fractal architecture,
fractal dimension, Frank Lloyd
Wright, Le Corbusier, Gothic
architecture, environmental
psychology, biophilia, biophilic
architecture

Research

Fractal Architecture Could Be Good for You

Abstract. The deployment of fractal principles in art and architecture seems to be a phenomenon of all times, and is in no way restricted to the period after the systematic mathematical understanding and description of fractals from the 1970s onwards. Nowadays, computer-generated fractal art, and the software to generate it, are widely available on the Internet. Fractal principles are also at work in more "traditional" arts or crafts, such as some Dalí paintings, mandalas, mosaics, floor decorations, and so on. This paper presents some of the architectural appropriations of fractal geometry. The concluding sections argue that fractal architecture is in a sense "good" for us.

The architectural utility of fractal geometry

Before I embark on a review of fractal forms in architecture, you should know that this type of geometry is used in architecture for two main reasons. First, some scholars have promoted fractal geometry as a creative tool. For example, Carl Bovill [1996] uses fractal rhythms, created by midpoint displacement, to generate a wide range of architectural organizations, such as planning grids, strip windows, noise abatements, and so on. Nikos Salingaros [1998] has emphasized three-dimensional applications of fractals in architecture and fervently argues for the recurrence of self-similar architectural elements on different scales of the built form. From a study of natural entities, he concludes that the scaling relationship between these elements should obey the ratio 2.7 to be aesthetically pleasing. Finally, note that an intuitive understanding of the creative value of fractal geometry was already present in classical architectural composition, which has lead to some remarkable similarities with well-known fractals (fig. 1). Andrew Crompton [2002] has tentatively argued that this is because some classical composition rules favour fractal forms.

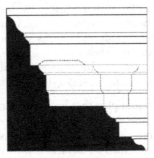

Fig. 1. A classical Doric entablature (left) has a remarkable similarity with the Devil's Staircase[1] (right). Picture credits: Andrew Crompton

A second way in which fractal geometry can be related to architecture is to use some of its typical measurement techniques to analyse the structure of buildings. Many readers are perhaps familiar with the box counting dimension, which is a measure for the recursiveness of detail on ever smaller scales.[2] Carl Bovill [1996] has applied this method to different building styles. He found that Wright's organic architecture shows a 'cascade of detail' on different scales, while in Le Corbusier's modernist architecture, the box counting dimension quickly drops to 1 for smaller scales. This finding is consistent with the fact that 'Wright's organic architecture called for materials to be used in a way that captured nature's complexity and order ... [while] Le Corbusier's purism called for materials to be used in a more industrial way, always looking for efficiency and purity of use' [Bovill 1996, 143]. Similar to Bovill, Daniele Capo [2004] applied the box counting method to the classical orders and found that there is detail up to 1/256th of the height of the entire order. Burkle-Elizondo and Valdéz-Cepeda [2006] also used fractal measurement techniques to establish the complexity of thirty-five Mesoamerican pyramids, and found that the monuments had a fractal dimension of around 1.3.

Two-dimensional fractals in architecture

Let me now show how fractal forms are, and have been, integrated in architecture. On first sight there does not seem to be an all-encompassing factor that binds the following buildings together. Sometimes, the fractal form is an expression of a worldview or a social idea, while on other occasions the architect just found it an attractive shape. Nevertheless, in the final sections I tentatively propose that there is perhaps a deeper-lying reason why such patterns are integrated in architecture, throughout all ages and cultures.

I start off with an overview of two-dimensional fractal forms in architecture, which are mostly present in the ground plans of buildings. You can find this application in a wide range of architectural structures, ranging from the plans of fortifications, to the organization of traditional Ba-ilia villages (Zambia). The global form of the latter settlements reoccurs in the family ring, which consists of individual houses, which are, again, similar to the overall shape of the village.

Interestingly, the scaling hierarchies governing this whole are a reflection of the social hierarchy in these communities [Eglash and Odumosu 2005]. As is noted by George Hersey [1993], a fractal organization is also characteristic of the plan of Bramante's design for St. Peter's in Rome:

> Symmetrically clustered within the inside corners formed by the cross's arms are four miniature Greek crosses, that, together, make up the basic cube of the church's body. The arms of these smaller crosses consist of further miniatures. And their corners, in turn, are filled in with smaller chapels and niches. In other words, Bramante's plan ... may be called fractal: it repeats like units at different scales [Hersey 1993].

The fractal ground plan that has perhaps received most theoretical attention is Wright's Palmer House (Ann Arbor, Michigan). In order to understand its fractal character, it is important to note that architects sometimes use a 'module' as the main organizational element. In a sense, such an element can be understood as the conceptual 'building block' of the house (e.g., a circle). Wright often applied this procedure to his work. Initially, the geometry governing his architecture created with the aid of such modules remained Euclidean. In later works, however, these elements were sometimes so organized that they gave the building a remarkable fractal organization. The Palmer House seems to be the culmination point of this evolution. Here, one geometric module – an equilateral triangle –

is repeated in the ground plan on no less than 7 different scales [Eaton 1998] (fig. 2). Another Wright building, whose fractal nature is visible – but in elevation – is the Town Hall in Marin County (San Francisco). In this structure, above each arch a window or arch is placed that is somewhat smaller than the previous one. This gives the structure self-similarity up to five scales [Portoghesi 2000].

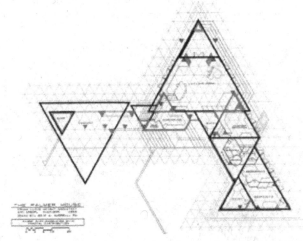

Fig. 2. Ground plan of Wright's Palmer House. Drawing by Eric Murrell, from [Eaton 1998].
Reproduced with permission, Kim Williams Books

Three-dimensional fractals in architecture

An obvious disadvantage of fractal ground plans is that the fractal component is barely visible for the viewer in a normal architectural experience. In this sense, it could be claimed that it loses some of its significance, and that three-dimensional applications are more convincing. A three-dimensional method, which some have linked to fractals, is to tessellate the architectural façade [Jencks 2002]. On first sight, the link with fractal geometry could seem obvious: such patterns are rich in detail, which is an intrinsic characteristic of fractals.

The contemporary architectural group Ashton Raggatt McDougall was perhaps one of the first to apply fractal tiling to architecture [Jencks 2002]. They covered the façade and the interior of Storey Hall (Melbourne) with polygon tiles that are inspired on Penrose tiling. Penrose tilings were discovered by the British mathematical physicist Roger Penrose and consist of a small number of simple tiles that can cover the plane in a non-repetitive manner. Such tessellations can be related to fractal geometry because they can be generated by Iterated Function Systems and L-systems. A 'tiling-approach' has also been adopted by the Lab Architecture Studio for Federation Square in Melbourne and its adjacent buildings. The main units of this 'fractally incremental system' are triangles, which are organized by five into panels, while five panels form the main constructional module (fig. 3).

I find the relation of tiled façades with fractal geometry difficult to judge. While some scholars, such as Charles Jencks, have unambiguously related such creations with fractals, it can also be observed that such constructs have no detail within detail, no tiles within tiles.

In a sense, the patterns are no more fractal than a grid of squares is fractal. They contain many details, but zooming in on the structure does not reveal new detail. On the other hand, in the case of Storey Hall and Federation Square, the tiles are organized into 'higher-order' wholes, with the aid of texture, colour and lines. This gives them a profounder sense of self-similarity. Finally, some might argue that such fractal interventions are merely surface treatment: essentially, these are not architectural but decorative interventions. Indeed, leaving out the patterns would probably wipe out the fractal character of the building altogether.

Fig. 3. Federation Square, Melbourne (Lab Architecture Studio). Picture credits: Steven Connor

But do there exist instances of modern architecture where the fractal component is eminently three-dimensional – where it pertains to the architectural form and/or structure? Such appropriations seem rare. On the Internet I came along the website "Fractal Architecture" (http://www.fractalarchitect.com), which shows building designs that are the result of marrying fractal principles and modernist forms. In the twentieth century Russian artist Malevich has created a series of architectural designs (*Arkhitektoiniki*) of which some have a remarkable fractal component. In one example, the main architectural form is surrounded by smaller versions of the whole building, which are again surrounded by even smaller fragments. The relation between their number and size is claimed to obey a $1/f$ relation. More recently, Steven Holl Architects' Simmons Hall has been related to fractal geometry, because it is inspired by a sponge whose openings have a fractal distribution.

For other eminent examples of three-dimensional fractal architecture, you have to go back in time. Sometimes it is noted that some of Leonardo Da Vinci's cathedral designs are fractal, because the domes are repeated for different sizes. However, this example (and the previous ones) cannot meet up to the profound fractal character of certain Hindu temples

[Trivedi 1989; Portoghesi 2000]. The fractal character of Hindu temples is strongly intertwined with Hindu cosmology (fig. 4).

Fig. 4. Fractal generation of central dome of a Hindu Temple

In fact, these edifices should be understood as models of the Hindu cosmos. In Hindu philosophy the cosmos is (more or less) conceived as a hologram, where each part of the whole is the whole itself, and contains all the information about the whole. Some schools of Hindu thought adhere to the (related) view that the macrocosm is 'enclosed' in the microcosm:

> The entire cosmos can be visualized to be contained in a microcosmic capsule, with the help of the concept of subtle elements called 'tanmatras'. The whole cosmic principle replicates itself again and again in ever smaller scales. The human being is said to contain within itself the entire cosmos [Trivedi 1989, 245-246].

Interestingly, both cosmological conceptions can be straightforwardly related to fractal self-similarity. Here also, the global structure recurs – over and over again – in the microstructure.

In order to maintain a harmonious worldview, man-made objects and artistic expressions were made in accordance with the central principles governing the Hindu cosmos – the result being a profound three-dimensional fractal architecture. The fractality of Hindu temples can be traced back to a set of typical architectural interventions. A survey of these methods is not only theoretically interesting, but also offers a set of concrete guidelines for enhancing the fractal character of architecture.

1. Splitting or breaking up a form, and repeating it horizontally, vertically or radially.
2. In the ground plan, iteratively replacing plain sides by sides that contain interior and exterior projections, or more detail. This method can also be applied in three dimensions.
3. Three-dimensional self-similar iteration of the central spire of a Hindu temple.
4. Repeating similar shapes horizontally and/or vertically.
5. Three-dimensional superimposition of architectural elements ('... motifs are inscribed within different kinds of motifs and several different kinds of themes and motifs are condensed and juxtaposed together into one complex and new entity' [Trivedi 1989, 257].)

Fig. 5. Notre Dame Cathedral, Paris

Fig. 6. Gothic architecture (City Hall, Bruges). The arched form reoccurs throughout the building façade on different hierarchical scales

In the West, the fractal Hindu temple seems to have its counterpart in Gothic architecture (figs. 5 and 6). From the illustrations you can easily see which specific methods are at the root of the fractality of the Gothic. In fact, these methods are quite similar to the ones used in Hindu architecture. For example, the shape of the main portal in fig. 5 recurs on a smaller scale in the two side portals. Further and smaller versions of this form can be found in the different arched windows or openings. In some Gothic buildings the windows are divided in constituent parts, where the stained glass, with its vibrant and colourful patterns, adds even more detail to the façade. The complexity of Gothic façades is further increased by the mere repetition of forms. For example, in fig. 5 the contours of the main and side portals are repeated inwardly, and circumscribed with a wealth of sculpted figures. The fractal component of Gothic cathedrals is also evident in the rose windows. These often contain flower-like patterns of varying sizes, and the stained glass often enhances the fractality even further.

Why fractal architecture could be good for you

Perhaps the last examples of fractal architecture are the most appealing to mathematicians, because the fractal component is clearly visible. However, as a philosopher it strikes me that these buildings have a strong aesthetic 'pull'. I can testify to this first-hand because I have the privilege of living in the medieval town of Bruges (fig. 8). Loads of tourists are fascinated by the Gothic buildings, and are ever so keen to take some snapshots, at the risk of being run over by the typical horse carriages that go around here. Of course there is an interplay of different factors here: people also go to see these monuments because the tourist industry thinks that they are worth seeing. Yet, I am certain that the mathematical – i.e., the fractal – structure of the buildings plays a role too. Some years ago I had the chance to embark on a research project that tried to address this issue. In essence, the research has been strongly interdisciplinary, and brings together findings from the fields of art, mathematics and psychology.

As strange as it may sound, I found that there is reason to believe that we are in a sense attuned to fractals, and hence, to fractal architecture. One line of evidence shows that the human nervous system is governed by time fractals. More specifically, analyses of brain functioning shows that it displays typical noise signals, commonly referred to as '1/f noise' or 'pink' noise [Anderson and Mandell 1996]. The fractal character of such noise can be easily appreciated, because like spatial fractals, it shows self-similar detail when you zoom in on it. But what is the function of these time fractals? A common answer is that the natural world – and the way it changes over time – is also characterized by pink noise, which suggests that our fractal minds are optimized to process the fractal characteristics of natural scenes (see, for example, [Knill et al. 1990]). This hypothesis is supported by the finding that discriminating fractal contours is best for those that have the same (fractal) properties as natural scenes [Gilden et al. 1993]. Interestingly, these findings can (tentatively) explain the creation of fractal artwork, and fractal architecture in particular. Such art should be understood as an exteriorization of the fractal aspects of brain functioning [Goldberger 1996]. Or as Goldberger [1996] puts it: '... the artwork externalizes and maps the internal brain-work ... Conversely, the interaction of the viewer with the artform may be taken as an act of self-recognition'.

However, as plausible as this explanation might sound, I don't find it satisfactory. The main reason is that it cannot account for why fractal patterns have a strong aesthetic component. A factual description of the human perceptual system in terms of time fractals does not explain why we find fractal structure beautiful. For a more plausible explanation for the aesthetic appeal of fractal architecture we have to delve into some findings from the field of environmental psychology (for a related account, see [Salingaros 2006]). Here, the empirical literature has shown, over and over again, that humans display a positive emotional affiliation with a specific set of natural elements and settings, namely vegetation (trees, plants and flowers) and savannas (see [Ulrich 1993] for an excellent review). More specifically, researchers found that these elements (a) lead to positive aesthetic responses, and (b) reduce psychological and physiological stress in humans [Ulrich et al. 1991]. These emotional affiliations are claimed to be 'hardwired' in the human brain and are sometimes framed in terms of "biophilia" (literally, "love for life"; see e.g., [Kellert and Wilson, 1993]). They are remnants of our shared human evolution in East-African savannas. Having these inborn "biofilic" responses was advantageous because they motivated our ancestors to explore and approach the natural settings and the elements it contained. This increased survival chances: vegetative elements were eminent sources of food, and could offer protection, while savannas are known to be high-quality habitats. People will also feel more relaxed in places that are good for living, hence the stress-reducing effect of such settings and of their constituent (vegetative) elements [Ulrich 1993].

At this point, the reader interested in mathematics might ask what this all has to do with fractals. Let's make my point. While the fractality of nature has been amply demonstrated, there is now reason to believe that the presence of fractal geometry (in a sense) underlies these biophilic responses. To put it very crudely, it is not the tree that the causes these emotional responses, but the fractal mathematics of the tree. Preliminary research that supports this hypothesis has been conducted by Caroline Hägerhäll and colleagues [2004]. They found that the emotional states towards vegetated/natural landscapes can be predicted by typical fractal characteristics (i.e., the fractal dimension). This study, and others, also indicate that the aesthetic reactions peak when the natural settings, or the fractal pattern underlying it, have an intermediate fractal dimension (around 1.3 – 1.5) [Aks and Sprott 1996; Abraham et al. 2003; Spehar et al. 2003]. What is even more surprising is that images with this range of fractal dimension also dampen stress in humans [Wise and Taylor 2002]. I, and others, have speculated that this reflects our inborn emotional affiliation with savanna-type landscapes, which are intermediately complex environments [Joye 2007]. In fact, I believe that the brain constantly evaluates settings with regard to their habitability. It would be adaptive if it could calculate the fractal dimension of a setting: this quickly conveys basic information about the complexity of a particular setting, which is a strong indicator whether it is a good place for living. Settings with a high fractal dimension could contain hidden dangers, such as ambushing predators, while those with a low fractal dimension probably do not contain enough elements, in order to offer protection and sources of food. Hence, we will be more aesthetically attracted and more relaxed in environments with an intermediate fractal dimension.

Now what is the link with architecture? In essence, the previous argument points out that, as a result of evolution, the brain has a preference for fractal structures, and feels relaxed when surrounded by these. This means that one of the reasons why we like the fractals in Gothic and Hindu architecture is that they remind us of our ancient, natural habitats. Because our brains have not fundamentally changed since prehistory, these biofilic responses are still at work. However, in the modern world, the fractal forms (e.g. trees) which we crave for are increasingly pushed back from our current habitats, and often replaced by simple Euclidean forms. Some scholars argue that this discrepancy can lead to an increased release of stress-related hormones, which, in the long run, can have deleterious effects on human health [Parsons 1991]. Others think that this discrepancy could be one of the underlying causes for psychopathologies in westernized societies [Gullone 1999]. With all this I do not claim that Euclidean buildings are inherently wrong or unhealthy; in fact, we are as much cultural as biological beings. However, the dominance of Euclidean shapes could prove harmful, and seems at variance with some of the mathematical preferences of our brains. As we all intend to live good lives, it wouldn't be a bad idea to replace some of them by architectural work that implements some essential fractal characteristics. Our preference for a specific fractal dimension further indicates that the effects of such buildings can be maximized for intermediate levels of complexity. So, the next time you feel stressed from a hard day's work, stimulate your brain with some fractals – albeit natural or architectural ones.

Notes

1. You construct this 'borderline' fractal as follows. Consider a square whose side length is 1. In the first step, make a column on the middle third part of the square with length 1/3 and height 1/2. In the second step, erect a column of height 1/4 over the interval 1/9 - 2/9 and one of height 3/4 over the interval 7/9 - 8/9. In the third step make four columns of heights 1/8, 3/8, 5/8, 7/8. For k steps 2 k-1 columns are drawn of heights 1/2k, 3/2k,... , (2k–1)/2k.

2. You obtain the box counting dimension as follows. First, place a rectangular grid over a (two-dimensional) representation of the architectural object. Count the number of boxes across the bottom of the grid (B1), and the number of boxes that contain a portion of the representation (N1). Next, decrease the size of the boxes of the grid, and again count the number of them at the bottom of the grid (B2), and the number that contains a fragment of the object (N2). Finally, plot a log (B) versus log (1/N) on a log-log diagram. The slope of this (straight) line approximates the box counting dimension.

References

ABRAHAM, F. D., et al. 2003. Judgements of time, aesthetics, and complexity as a function of the fractal dimension of images formed by chaotic attractors. http://www.blueberry-brain.org/silliman/JEM%20ms2.htm.

AKS, D.J. and J. C. SPROTT. 1996. Quantifying aesthtic preference for chaotic patterns. *Empirical studies of the arts* **14**: 1-6.

ANDERSON, C.M. and A.J. MANDELL. 1996. Fractal time and the foundations of consciousness: vertical convergence of 1/f phenomena from ion channels to behavioural states Pp. 75-126 in *Fractals of brain, fractals of mind*, E. MacCormac and M.I. Stamenov eds. Amsterdam/Philadelphia: John Benjamins.

BOVILL, C. 1996. *Fractal Geometry in Architecture and Design*. Boston: Birkhäuser.

BURKLE-ELIZONDO, G. and VALDÉZ-CEPEDA, R.D. 2006. Fractal Analysis of Mesoamerican Pyramids. *Nonlinear Dynamics, Psychology, and Life Sciences* **10**: 105-122.

CAPO, D. 2004. The Fractal Nature of the Architectural Orders. *Nexus Network Journal* 6, 1: 30-40.

CROMPTON, A. 2002. Fractals and picturesque composition *Environment and Planning B: Planning and Design* **29**: 451-459.

EATON, L. 1998. Fractal Geometry in the Late Works of Frank Lloyd Wright. Pp. 23-38 in *Nexus II: Architecture and Mathematics*, K. Williams, ed. Fucecchio (Florence): Edizioni Dell'Erba.

EGLASH, P. and T.B. ODUMOSU. 2005. Pp. 101- 109 in *What Mathematics from Africa?*, G. Sica (ed.) Monza: Polimetrica International Scientific Publisher.

GOLDBERGER, A.L. 1996. Fractals and the birth of Gothic: reflections on the biological basis of creativity. *Molecular Psychiatry* **1**: 99-104.

GILDEN, D.L., et al. 1993. *Psychological Review* 100: 460-78.

GULLONE, E. 2000. The biophilia hypothesis and life in the 21st century: increasing mental health or increasing pathology. *Journal of Happiness Studies* 1: 293-321.

HÄGERHÄLL, C., et al. 2004. Fractal dimension of landscape silhousette outlines as a predictor of landscape preference. *Journal of Environmental Psychology* 24: 247-255.

HERSEY, G. 1993. *High Renaissance Art in St. Peter's and the Vatican: An Interpretive Guide.* Chicago: University of Chicago Press.

JENCKS, C. 2002. *The New Paradigm in Architecture.* New Haven & London: Yale University Press.

JOYE, Y. In preparation. Towards Nature-Based Architecture: Drawing Lessons from Psychology. *Review of General Psychology* (2007).

KNILL, D.C., et al. 1990. Human discrimination of fractal images. *Journal of the Optical Society of America A* 7: 1113-1123.

PARSONS, R. 1991. The potential influences of environmental perception on human health. *Journal of Environmental Psychology* 11: 1-23.

PORTOGHESI, P. 2000. *Nature and Architecture.* Milan: Skira.

SALINGAROS, N. 1998. A scientific basis for creating architectural forms. *Journal of Architectural and Planning Research* 15: 283-293.

SPEHAR, B., et al. 2003. Universal aesthetic of fractals. *Computers & Graphics* 27: 813-820.

TRIVEDI, K. 1988. Hindu temples: models of a fractal universe. *Space design* 290: 234-258.

ULRICH, R.S., et al. 1991. Stress recovery during exposure to natural and urban environments. *Journal of Environmental Psychology* 11: 201-230.

ULRICH, R.S. 1993. Biophilia, biophobia, and natural landscapes. Pp. 73-137 in *The Biophilia Hypothesis*, S.R. Kellert and E.O. Wilson, eds. Washington: Island Press.

WISE, J.A. and R. TAYLOR. 2002. Fractal design strategies for enhancement of knowledge work environments. Pp. 854-859 in *Proceedings of the 46th meeting of the Human Factors and Ergonomics Society.* Santa Monica, CA: Human Factors and Ergonomics Society.

About the author

Yannick Joye obtained his Ph.D. in philosophy at the University of Ghent, Belgium. His research focuses on the question which architectural interventions can be deduced from the fact that humans have evolved in natural, fractal-like habitats. He is thereby among those that try to provide a theoretical framework for the upcoming field of biophilic architecture.

Antonia Redondo Buitrago

I.E.S. Bachiller Sabuco
Departamento de Matemáticas.
Avenida de España 9
02002 Albacete SPAIN
amcredondo@terra.es

Keywords: Bronze mean, diagonals, metallic means, polygons

Research

Polygons, Diagonals, and the Bronze Mean

Abstract. This article furthers the study of the Metallic Means and investigates the question of whether or not there exists a polygon corresponding to the Bronze Mean as the pentagon and the octagon correspond respectively to the Golden and Silver Means.

The Metallic Means Family (MMF) was introduced in 1998 by Vera W. de Spinadel [1998; 1999]. Its members are the positive solutions of the quadratic equations $x^2 - px - q = 0$, where the parameters p and q are positive integer numbers. The more relevant of them are the Golden Mean and the Silver Mean, the first members of the subfamily which is obtained by considering $q = 1$. The members σ_p, $p = 1,2,3K$ of this subfamily share properties which are the generalization of the Golden Mean properties. For instance, they all may be obtained by the limit of consecutive terms of certain "*generalized secondary Fibonacci sequences*" (GSFS) and they are the only numbers which yield geometric sequences:

$$K \, , \frac{1}{\left(\sigma_p\right)^3}, \frac{1}{\left(\sigma_p\right)^2}, \frac{1}{\sigma_p}, 1, \sigma_p, \left(\sigma_p\right)^2, \left(\sigma_p\right)^3, K$$

with additive properties

$$1 + p\sigma_p = \left(\sigma_p\right)^2, \quad \left(\sigma_p\right)^k + p\left(\sigma_p\right)^{k+1} = \left(\sigma_p\right)^{k+2}$$

$$\frac{1}{\left(\sigma_p\right)^k} = \frac{p}{\left(\sigma_p\right)^{k+1}} + \frac{1}{\left(\sigma_p\right)^{k+2}}, \quad k = 1,2,3,K \, .$$

However, the generalization of geometrical aspects presents some differences. If we consider the rectangles of ratio σ_p, they all have the property that the corresponding gnomon is the union of p squares. Figs. 1, 2 and 3 show the three first metallic rectangles.

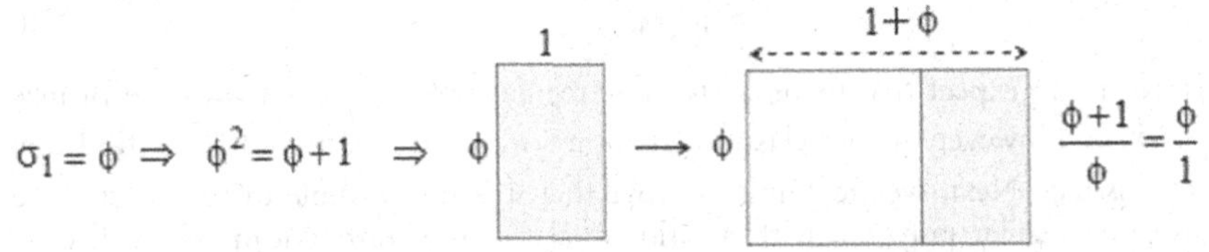

$$\sigma_1 = \phi \implies \phi^2 = \phi + 1 \implies$$

Fig. 1

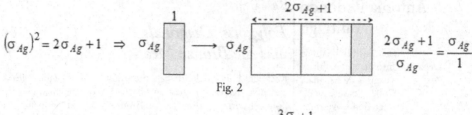

$$\left(\sigma_{Ag}\right)^2 = 2\sigma_{Ag} + 1 \implies \sigma_{Ag} \qquad \frac{2\sigma_{Ag}+1}{\sigma_{Ag}} = \frac{\sigma_{Ag}}{1}$$

Fig. 2

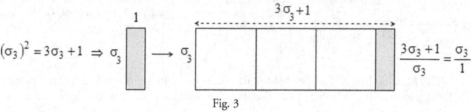

$$\left(\sigma_3\right)^2 = 3\sigma_3 + 1 \implies \sigma_3 \qquad \frac{3\sigma_3+1}{\sigma_3} = \frac{\sigma_3}{1}$$

Fig. 3

Also, the Silver Mean, $\sigma_{Ag} = \sigma_2$, and the Bronze Mean, $\sigma_{Br} = \sigma_3$, allow us to construct spirals which generalize to that of the Golden Mean [Redondo Buitrago 2006].

On the other hand, the Golden Mean is linked to pentagonal symmetry, and the Silver Mean to octagonal symmetry. Actually, in the regular pentagon, the ratio of the lengths of the first diagonal to that of the side is $\sigma_1 = \phi$, and the Silver Mean, $\sigma_2 = \sigma_{Ag}$, is the ratio of the lengths of the second diagonal to that of the side in the regular octagon.

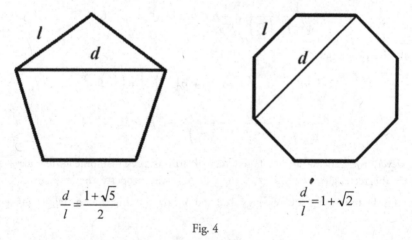

$$\frac{d}{l} = \frac{1+\sqrt{5}}{2} \qquad\qquad \frac{d}{l} = 1+\sqrt{2}$$

Fig. 4

So, it is natural to expect that there exists some regular polygon linked with the Bronze Mean $\sigma_3 = \sigma_{Br}$. However, in the classical literature we have not been able to find any reference to this fact. Next, we are going to prove that it is not possible to construct some diagonal in some regular polygon, with a ratio equal to the Bronze Mean. We will only need very elementary geometrical arguments.

Let there be a regular polygon with n sides. When we draw and number its diagonals d_1, d_2,..., d_{n-1}, including the sides of the polygon as d_1 and d_{n-1}, as in Figs. 5 and 6, we observe that the length of d_i is equal to d_{n-1}.

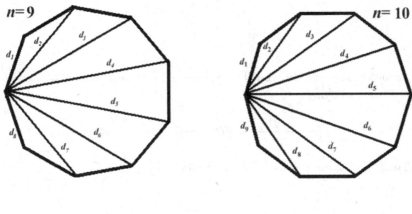

Fig. 5   Fig. 6

So, we are going to consider only the lengths of diagonals d_1, d_2,..., d_{n_0}, where n_0 stands for the integer part of $(n-2):2$. Obviously, if n is an even number, as in fig. 6, then the largest diagonal coincides with the diameter of the circumscribed circumference.

The ratios of the lengths of the diagonals are given by the law of cosines (fig. 7).

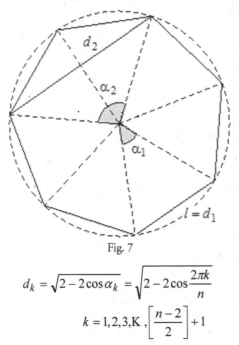

Fig. 7

$$d_k = \sqrt{2 - 2\cos\alpha_k} = \sqrt{2 - 2\cos\frac{2\pi k}{n}}$$

$$k = 1,2,3,\text{K}, \left[\frac{n-2}{2}\right] + 1$$

In particular, taking into account that the side of the polygon is d_1, we have

$$\frac{d_k}{l} = \sqrt{\frac{1-\cos\dfrac{2\pi k}{n}}{1-\cos\dfrac{2\pi}{n}}} = \frac{\sin\dfrac{\pi k}{n}}{\sin\dfrac{\pi}{n}}.$$

So, in order to research the possibility of existence of diagonals in some regular polygon with $n \geq 5$ sides, satisfying

$$\frac{d_k}{l} = \frac{3+\sqrt{13}}{2} = 3.30277563K \quad , \quad k = 2,3,K, \left[\frac{n-2}{2}\right]+1$$

we will study the following periodic function family

$$f_n(x) = \frac{\sin\dfrac{\pi x}{n}}{\sin\dfrac{\pi}{n}}, \; n \geq 5$$

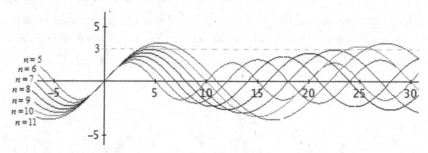

Fig. 8

First, we observe that all functions have period $T = 2n$. Moreover, in $[0, 2n]$ the maxim value of the function is $f(n/2) = (\sin(\pi/n))^{-1}$ (fig. 9).

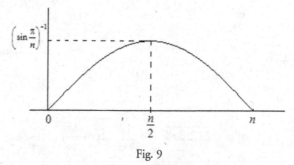

Fig. 9

Notice that when the number of sides, n, is fixed by computing the values of the corresponding function at the integer values of the interval $[0, n]$, we obtain the ratio of the successive lengths of the side and diagonals of the corresponding polygon. Nevertheless, only we need consider the first half of them. That is, the interval $[0, n/2]$.

If we take a look at the function graph above, we observe that the maximum for $n<10$ is less than 3. Therefore, we must search out polygons with $n\geq10$. But, by checking $n =10$, 11, 12, 13 y 14, we deduce that it is not possible to get

$$f_n(k) = \frac{3+\sqrt{13}}{2} = 3.30277563K \quad , \quad k = 2,3,K, \left[\frac{n-2}{2}\right]+1$$

n	10	11	12	13	14
$f_n(k)$	3.236 (k=5)	3.229 (k=4)	2.732 (k=3)	2.771 (k=3)	2.802 (k=3)
		3.513 (k=5)	3.334 (k=4)	3.439 (k=4)	3.513 (k=4)
			3.732 (k=5)	3.907 (k=5)	4.049 (k=5)
			3.863 (k=6)	4.148 (k=6)	4.381 (k=6)
					4.494 (k=7)

On the other hand, when k is fixed, the function $g_k(n) = f_n(k)$ is an increasing function, and fig. 10 shows that we need to study only the values 2, 3 and 4, that is the second, third and fourth diagonals.

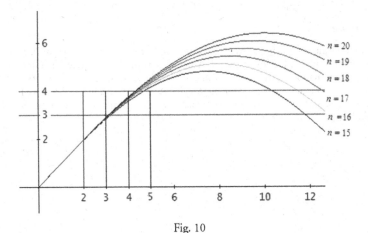

Fig. 10

Hence, we find that

$$1.95... = f_{15}(2) < f_{16}(2) < f_{17}(2) < \Lambda < \lim_{n \to +\infty} \frac{\sin\left(\dfrac{2\pi}{n}\right)}{\sin\left(\dfrac{\pi}{n}\right)} = 2 < \frac{3+\sqrt{13}}{2}$$

$$2.82... = f_{15}(3) < f_{16}(3) < f_{17}(3) < \Lambda < \lim_{n \to +\infty} \frac{\sin\left(\dfrac{3\pi}{n}\right)}{\sin\left(\dfrac{\pi}{n}\right)} = 3 < \frac{3+\sqrt{13}}{2}$$

$$\frac{3+\sqrt{13}}{2} < 3.57... = f_{15}(4) < f_{16}(4) < f_{17}(4) < \Lambda < \lim_{n \to +\infty} \frac{\sin\left(\dfrac{4\pi}{n}\right)}{\sin\left(\dfrac{\pi}{n}\right)} = 4$$

Consequently, we can conclude: There exists no regular polygon in which the ratio of the length of the diagonal to the side of the polygon is equal to the Bronze Mean.

References

REDONDO BUITRAGO, Antonia. 2006. Algunos resultados sobre Números Metálicos. *Journal of Mathematics & Design* **6**, 1: 29-45.
SPINADEL, Vera W. de. 1998. *From the Golden Mean to Chaos.* 2nd ed. Buenos Aires: Nueva Librería
———. 1999. The family of metallic means. *Visual Mathematics* **1**, 3. http//members.tripod.com/vismath1/spinadel/

About the author

Antonia Redondo Buitrago teaches Mathematics in a high school of Albacete (Spain). She is a doctor in Applied Mathematics by University of Valencia (Spain). Her research interests and her contributions in international journals and congresses include works about the fractional powers of operators, continued fractions and the Metallic Means. At the present, in the domain of mathematics and design, her collaborations with Vera W. Spinadel in the research of new properties of Metallic Number Family are the most relevant.

Rachel Fletcher

113 Division St.
Great Barrington, MA 01230
USA
rfletch@bcn.net

Keywords: descriptive geometry,
diagonal, dynamic symmetry,
incommensurate values,
Lindisfarne Gospels, root
rectangles

Geometer's Angle

Dynamic Root Rectangles
Part One: The Fundamentals

Abstract. Incommensurable ratios cannot be stated in finite, whole number fractions. But such ratios can organize spatial compositions so that the same ratio persists through endless divisions. We explore this proportioning principle, which Jay Hambidge calls "dynamic symmetry," as it appears in "root rectangles" of incommensurable proportions.

I Root Rectangles

Incommensurable ratios orchestrate diverse spatial elements through endless divisions, achieving unity without compromising the integrity of individual parts. "Dynamic symmetry" is the term given by Jay Hambidge to describe this proportioning principle which he observes in ancient Greek and Egyptian art, the human figure, and other forms of organic life.[1] Dynamic symmetry governs the relationship between individual elements and the relationship they bear to the whole. It is the "perfect modulating process," imparting rhythm and movement to the transition from one level of composition to the next.[2] We outline the fundamental components of Hambidge's system, looking to future columns to explore specific design applications.

Definitions:

"**Incommensurable**" is an adaptation of the medieval Latin *incommensurabilis* (*in-* "not" + *com-* "together" + *mensura* "a measure"). Quantities that lack a common measure or factor, expressed as never-ending fractions, are incommensurable [Simpson 1989].

Square root is the number that produces a specified number when multiplied by itself. The square root of 25 is 5, and is written $\sqrt{25} = 5$.

Root rectangles are rectangles based on square root proportions. Their edge lengths are incommensurable and cannot divide into one another. But a square constructed on the long side of the rectangle can be expressed in whole numbers, relative to a square constructed on the shorter unit side [Hambidge 1967, 18].

"Dynamic symmetry" appears in root rectangles whose side lengths in ratio are incommensurable, but are commensurable or measurable when expressed in square. [Hambidge 1960, 22-24; 1967, 17-18].[3] A square constructed on the long side of a root-two rectangle (1 : $\sqrt{2}$ or 1 : 1.4142...) is double the area of a square constructed on the shorter unit side (fig. 1a). A square constructed on the long side of a root-three rectangle (1:$\sqrt{3}$ or 1:1.732...) is three times the area of a square constructed on the shorter unit side (fig. 1b).

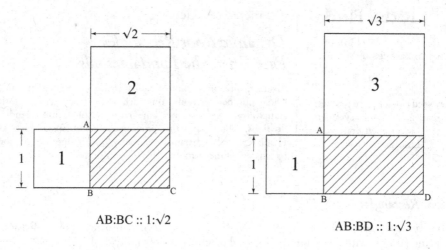

AB:BC :: 1:√2

Fig. 1a

AB:BD :: 1:√3

Fig. 1b

A square constructed on the long side of a root-four rectangle (1 : √4 or 1 : 2) is four times the area of a square constructed on the shorter unit side (fig. 1c).[4] A square constructed on the long side of a root-five rectangle (1 : √5 or 1 : 2.236...) is five times the area of a square constructed on the shorter unit side (fig. 1d).

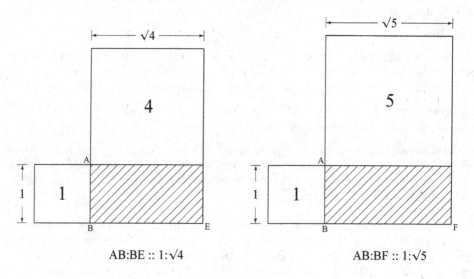

AB:BE :: 1:√4

Fig. 1c

AB:BF :: 1:√5

Fig. 1d

II Root Rectangles Outside a Square

A series of expanding root rectangles can be generated from a square, where each diagonal of the square or rectangle equals the long side of the next larger four-sided figure.

- Draw a horizontal baseline AB equal in length to one unit.
- From point A, draw an indefinite line perpendicular to line AB that is slightly longer in length.
- Place the compass point at A. Draw a quarter-arc of radius AB that intersects the line AB at point B and the open-ended line at point C.
- Place the compass point at B. Draw a quarter-arc (or one slightly longer) of the same radius, as shown.
- Place the compass point at C. Draw a quarter-arc (or one slightly longer) of the same radius, as shown.
- Locate point D, where the two quarter-arcs (taken from points B and C) intersect.
- Place the compass point at D. Draw a quarter-arc of the same radius that intersects the indefinite vertical line at point C and the line AB at point B (fig. 2).
- Connect points D, B, A and C.

The result is a square (DBAC) of side 1 (fig. 3).

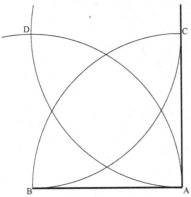

Fig. 2   Fig. 3

- Draw the diagonal BC through the square (DBAC).

The side (DB) and the diagonal (BC) are in the ratio $1 : \sqrt{2}$.[5]

- Place the compass point at B. Draw an arc of radius BC that intersects the extension of line BA at point E (fig. 4).

- From point E, draw a line perpendicular to line EB that intersects the extension of line DC at point F.
- Connect points D, B, E and F.

The result is a root-two rectangle (DBEF) with short and long sides in the ratio $1 : \sqrt{2}$ (fig. 5).

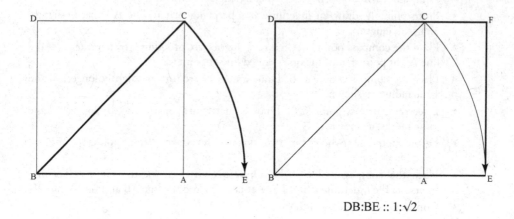

DB:BE :: 1:√2

Fig. 4   Fig. 5

- Draw the diagonal BF through the root-two rectangle (DBEF).

The side (DB) and the diagonal (BF) are in the ratio 1 : √3.

- Place the compass point at B. Draw an arc of radius BF that intersects the extension of line BE at point G (fig. 6).

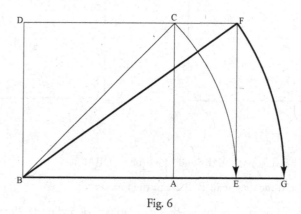

Fig. 6

- From point G, draw a line perpendicular to line GB that intersects the extension of line DF at point H.
- Connect points D, B, G and H.

The result is a root-three rectangle (DBGH) with short and long sides in the ratio 1 : √3 (fig. 7).

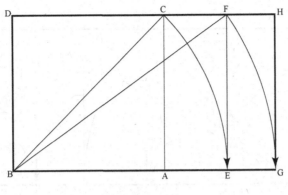

DB:BG :: 1:√3

Fig. 7

- Draw the diagonal BH through the root-three rectangle (DBGH).

The side (DB) and the diagonal (BH) are in the ratio 1 : √4.

- Place the compass point at B. Draw an arc of radius BH that intersects the extension of line BG at point I (fig. 8).

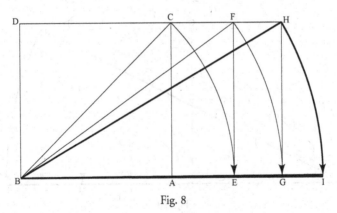

Fig. 8

- From point I, draw a line perpendicular to line IB that intersects the extension of line DH at point J.
- Connect points D, B, I and J.

The result is a root-four rectangle (DBIJ) with short and long sides in the ratio 1 : √4 (fig. 9).

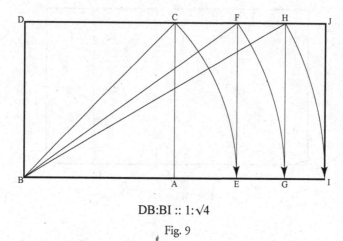

DB:BI :: 1:√4

Fig. 9

- Draw the diagonal BJ through the root-four rectangle (DBIJ).

The side (DB) and the diagonal (BJ) are in the ratio 1 : √5.

- Place the compass point at B. Draw an arc of radius BJ that intersects the extension of line BI at point K (fig. 10).

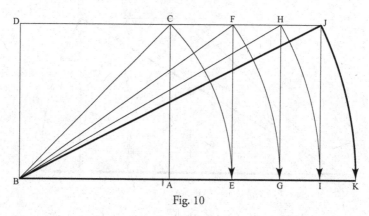

Fig. 10

- From point K, draw a line perpendicular to line KB that intersects the extension of line DJ at point L.

- Connect points D, B, K and L.

The result is a root-five rectangle (DBKL) with short and long sides in the ratio 1 : √5 (fig. 11).

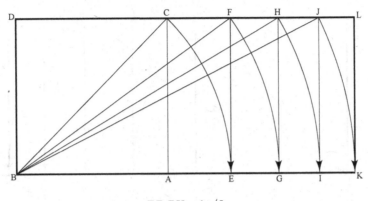

DB:BK :: 1:√5

Fig. 11

- Draw the diagonal BL through the root-five rectangle (DBKL).

The side (DB) and the diagonal (BL) are in the ratio 1 : √6.

- Place the compass point at B. Draw an arc of radius BL that intersects the extension of line BK at point M (fig. 12).

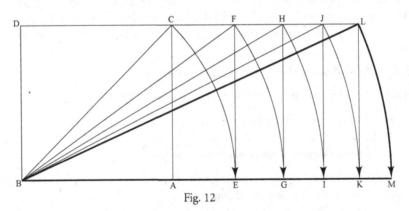

Fig. 12

- From point M, draw a line perpendicular to line MB that intersects the extension of line DL at point N.
- Connect points D, B, M and N.

The result is a root-six rectangle (DBMN) with short and long sides in the ratio 1 : √6 (fig. 13).

Fig. 13 displays the root rectangles in series, with their long sides relative to a unit side of 1.

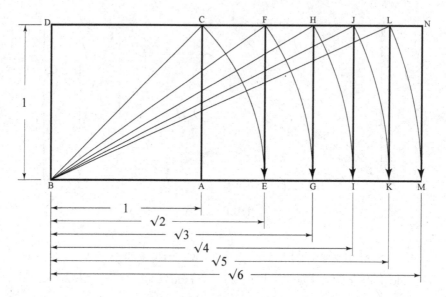

Fig. 13

Civil engineer A. E. Berriman observes ancient Egyptian units of measure in the square and the root-two and root-three rectangles. If the edge of the square equals 14.58 inches, or one Egyptian remen, the long edge of the 1 : √2 rectangle equals one royal cubit (20.625 inches); and the long edge of the 1 : √3 rectangle equals one Palestinian cubit (25.25 inches).[6]

III Square Root Numbers in Three Dimensions

Square root numbers can also be observed in three dimensions.

- Draw or construct a cube of side 1.
- Locate a diagonal of a square face.
- Locate a diagonal through the body of the cube.

If the edge of the cube equals 1, the diagonal of a square face equals √2, and the diagonal through the body of the cube equals √3 (fig. 14).

- To the original cube, add a second cube of equal size, as shown.
- Locate a long edge of the double cube.
- Locate a diagonal of a double square face.

If the short edge of the double cube equals 1, the long edge of a double square face equals √4, and the diagonal through a double square face equals √5 (fig. 15).

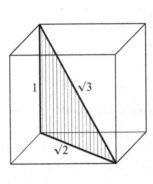

Fig. 14

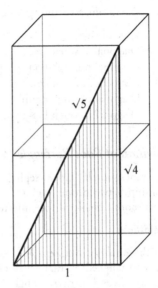

Fig. 15

- Locate a diagonal of a double square face.
- Locate a diagonal through the body of the double cube.

If the short edge of the double cube equals 1, the diagonal through a double square face equals √5, and the diagonal through the body of the double cube equals √6 (fig. 16).[7]

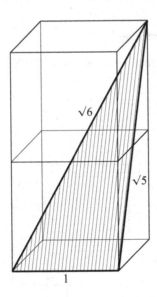

Fig. 16

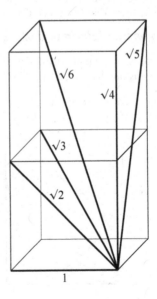

Fig. 17

Figure 17 summarizes the appearance of square root numbers in a double cube.

1	Edge of cube
√2	Diagonal of square face
√3	Diagonal through body of cube
√4	Long edge of double square face
√5	Diagonal through double square face
√6	Diagonal through body of double cube

IV Division of Root Rectangles into Reciprocals by the Diagonal

Incommensurable ratios replicate through endless spatial divisions and express the relationship between one level and the next. In Hambidge's system, this is evident in the way that root rectangles divide into reciprocals of the same proportion.

Definitions:

The **diagonal** is the straight line joining two nonadjacent vertices of a plane figure, or two vertices of a polyhedron that are not in the same face. The Greek for "diagonal" is *diagónios* (from *dia* "across" + *gónia* "angle"), which means "from angle to angle" [Liddell 1940, Simpson 1989].

The **reciprocal** of a major rectangle is a figure similar in shape, but smaller in size, such that the short side of the major rectangle equals the long side of the reciprocal [Hambidge 1967, 30, 131].

The diagonal of a major rectangle intersects the diagonal of its reciprocal at right angles in an arrangement that locates endless divisions in continued proportion.

V Diagonal and Reciprocal of the Root-Two Rectangle

A root-two rectangle divides into two reciprocals in the ratio 1 : √2. The area of each reciprocal is one-half the area of the whole.

- Draw a square (DBAC) of side 1. (Repeat figs. 2 and 3.)

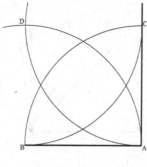

Fig. 2

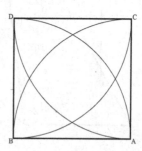

Fig. 3

- Draw a rectangle (DBEF) of sides 1 and √2, as shown. (Repeat figures 4 and 5.)

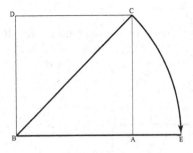

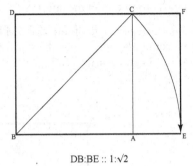

DB:BE :: 1:√2

Fig. 4   Fig. 5

- Locate the diagonal (BF) of the rectangle DBEF.
- Locate the line EF. Draw a semi-circle that intersects the diagonal BF at point O, as shown.
- From point E, draw a line through point O that intersects line FD at point G.
- From point G, draw a line perpendicular to line FD that intersects the line BE at point H.
- Connect points H, E, F and G.

The result is a smaller rectangle (HEFG) with short and long sides of 1/√2 and 1 (√2/2:1 or 0.7071... :1). Rectangle HEFG is the reciprocal of the whole rectangle DBEF.

The major 1 : √2 rectangle DBEF divides into two reciprocals (HEFG and BHGD) that are proportionally smaller in the ratio 1 : √2.

- Locate the diagonal (BF) of the rectangle DBEF.
- Locate the diagonal (EG) of the reciprocal HEFG.

The diagonals BF and EG intersect at 90° at point O (fig. 18).[8]

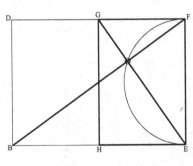

GF:FE :: FE:EB :: 1:√2
EG:BF :: 1:√2

Fig. 18

The diagonals EG and BF divide at point O into four radii vectors (OG, OF, OE and OB). The radii vectors emanate from the pole (point O) at equal 90° angles and progress in continued proportion in the ratio 1 : $\sqrt{2}$.[9]

A root-two rectangle of any size divides into two reciprocals in the ratio 1 : $\sqrt{2}$. If the process continues indefinitely, the side lengths of successively larger rectangles form a perfect geometric progression (1 , $\sqrt{2}$, 2, $2\sqrt{2}$...). The radii vectors locate where the divisions take place (fig. 19).

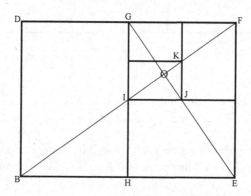

OG:OF :: OF:OE :: OE:OB :: 1:$\sqrt{2}$

KJ:JI :: JI:IG :: IG:GF :: GF:FE :: FE:EB :: 1:$\sqrt{2}$

Fig. 19

VI Diagonal and Reciprocal of the Root-Three Rectangle

A root-three rectangle divides into three reciprocals in the ratio 1 : $\sqrt{3}$. The area of each reciprocal is one-third the area of the whole.

- Draw a rectangle (DBGH) of sides 1 and $\sqrt{3}$, as shown. (Repeat figures 6 and 7.)

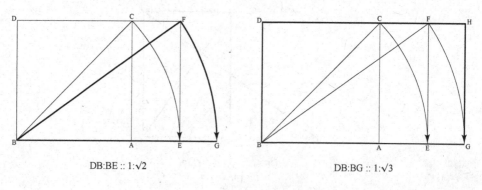

DB:BE :: 1:$\sqrt{2}$ DB:BG :: 1:$\sqrt{3}$

Fig. 6   Fig. 7

- Locate the diagonal (BH) of the rectangle DBGH.

- Locate the line GH. Draw a semi-circle that intersects the diagonal BH at point O, as shown.
- From point G, draw a line through point O that intersects line HD at point I.
- From point I, draw a line perpendicular to line HD that intersects the line BG at point J.
- Connect points J, G, H and I.

The result is a smaller rectangle (JGHI) with short and long sides of 1/√3 and 1 (√3/3:1 or 0.5773... :1). Rectangle JGHI is the reciprocal of the whole rectangle DBGH.

The major 1 : √3 rectangle DBGH divides into three reciprocals (JGHI, LJIK and BLKD) that are proportionally smaller in the ratio 1 : √3.

- Locate the diagonal (BH) of the rectangle DBGH.
- Locate the diagonal (GI) of the reciprocal JGHI.

The diagonals BH and GI intersect at 90° at point O (fig. 20).[10]

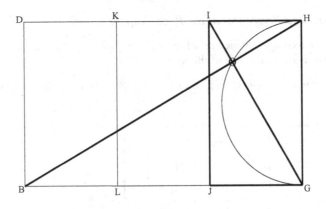

IH:HG :: HG:GB :: 1:√3

GI:BH :: 1:√3

Fig. 20

The diagonals GI and BH divide at point O into four radii vectors (OI, OH, OG and OB). The radii vectors emanate from the pole (point O) at equal 90° angles and progress in continued proportion in the ratio 1 : √3.[11]

A root-three rectangle of any size divides into three reciprocals in the ratio 1 : √3. If the process continues indefinitely, the side lengths of successively larger rectangles form a perfect geometric progression (1 ,√3, 3, 3√3...). The radii vectors locate where the divisions take place (fig. 21).

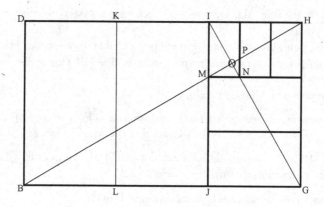

$$OI:OH :: OH:OG :: OG:OB :: 1:\sqrt{3}$$
$$PN:NM :: NM:MI :: MI:IH :: IH:HG :: HG:GB :: 1:\sqrt{3}$$

Fig. 21

VII Diagonal and Reciprocal of the Root-Four Rectangle

A root-four rectangle divides into four reciprocals in the ratio $1:\sqrt{4}$. The area of each reciprocal is one-fourth the area of the whole.

- Draw a rectangle (DBIJ) of sides 1 and $\sqrt{4}$, as shown. (Repeat figs. 8 and 9.)

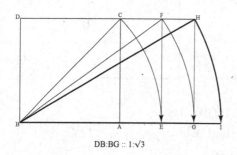

DB:BG :: $1:\sqrt{3}$ DB:BI :: $1:\sqrt{4}$

Fig. 8   Fig. 9

- Locate the diagonal (BJ) of the rectangle DBIJ.
- Locate the line IJ. Draw a semi-circle that intersects the diagonal BJ at point O, as shown.
- From point I, draw a line through point O that intersects the line JD at point K.
- From point K, draw a line perpendicular to line JD that intersects the line BI at point L.
- Connect points L, I, J and K.

The result is a smaller rectangle (LIJK) with short and long sides of 1/√4 and 1 (√4/4 : 1 or 0.5 : 1). Rectangle LIJK is the reciprocal of the whole rectangle DBIJ.

The major 1 : √4 rectangle DBIJ divides into four reciprocals (LIJK, NLKM, QNMP and BQPD) that are proportionally smaller in the ratio 1 : √4.

- Locate the diagonal (BJ) of the rectangle DBIJ.
- Locate the diagonal (IK) of the reciprocal LIJK.

The diagonals BJ and IK intersect at 90° at point O (fig. 22).[12]

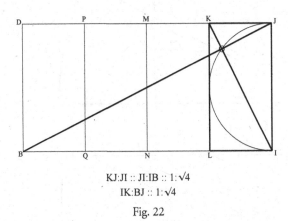

KJ:JI :: JI:IB :: 1:√4
IK:BJ :: 1:√4

Fig. 22

The diagonals IK and BJ divide at point O into four radii vectors (OK, OJ, OI and OB). The radii vectors emanate from the pole (point O) at equal 90° angles and progress in continued proportion in the ratio 1 : √4.

A root-four rectangle of any size divides into four reciprocals in the ratio 1 : √4. If the process continues indefinitely, the side lengths of successively larger rectangles form a perfect geometric progression (1 ,√4, 4, 4√4...). The radii vectors locate where the divisions take place (fig. 23).

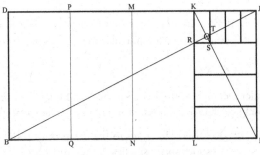

OK:OJ :: OJ:OI :: OI:OB :: 1:√4
TS:SR :: SR:RK :: RK:KJ :: KJ:JI :: JI:IB :: 1:√4

Fig. 23

VIII Diagonal and Reciprocal of the Root-Five Rectangle

A root-five rectangle divides into five reciprocals in the ratio 1 : √5. The area of each reciprocal is one-fifth the area of the whole.

- Draw a rectangle (DBKL) of sides 1 and √5, as shown. (Repeat figures 10 and 11.)

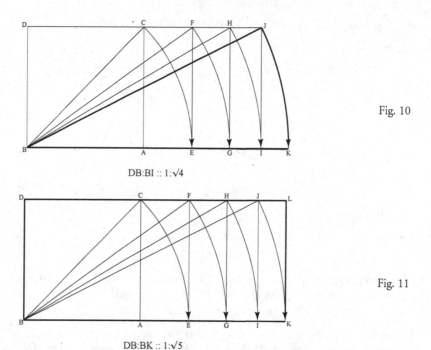

Fig. 10

DB:BI :: 1:√4

Fig. 11

DB:BK :: 1:√5

- Locate the diagonal (BL) of the rectangle DBKL.
- Locate the line KL. Draw a semi-circle that intersects the diagonal BL at point O, as shown.
- From point K, draw a line through point O that intersects line LD at point M.
- From point M, draw a line perpendicular to line LD that intersects the line BK at point N.
- Connect points N, K, L and M.

The result is a smaller rectangle (NKLM) with short and long sides of 1/√5 and 1 (√5/5:1 or 0.4472... :1). Rectangle NKLM is the reciprocal of the whole rectangle DBKL.

The major 1 : √5 rectangle DBKL divides into five reciprocals (NKLM, QNMP, SQPR, USRT and BUTD) that are proportionally smaller in the ratio 1 : √5.

- Locate the diagonal (BL) of the rectangle DBKL.
- Locate the diagonal (KM) of the reciprocal NKLM.

The diagonals BL and KM intersect at 90° at point O (fig. 24).[13]

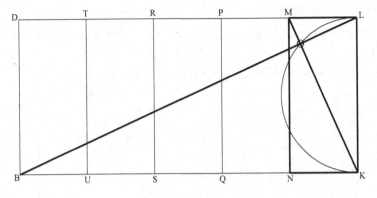

Fig. 24

$$ML:LK :: LK:KB :: 1:\sqrt{5}$$
$$KM:BL :: 1:\sqrt{5}$$

The diagonals KM and BL divide at point O into four radii vectors (OM, OL, OK and OB). The radii vectors emanate from the pole (point O) at equal 90° angles and progress in continued proportion in the ratio $1 : \sqrt{5}$.

A root-five rectangle of any size divides into five reciprocals in the ratio $1 : \sqrt{5}$. If the process continues indefinitely, the side lengths of successively larger rectangles form a perfect geometric progression ($1, \sqrt{5}, 5, 5\sqrt{5}...$). The radii vectors locate where the divisions take place (fig. 25).

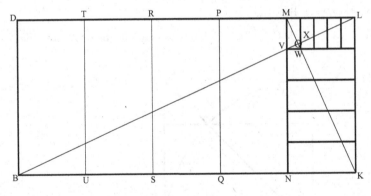

Fig. 25

$$OM:OL :: OL:OK :: OK:OB :: 1:\sqrt{5}$$
$$XW:WV :: WV:VM :: VM:ML :: ML:LK :: LK:KB :: 1:\sqrt{5}$$

A root-two rectangle equals 1 : 1.4142 ... . Its reciprocal ($\sqrt{2}/2$: 1) equals 0.7071... : 1.
A root-three rectangle equals 1 : 1.732 ... . Its reciprocal ($\sqrt{3}/3$: 1) equals 0.5773... : 1.
A root-four rectangle equals 1 : 2.0.   Its reciprocal ($\sqrt{4}/4$: 1) equals 0.5 : 1.
A root-five rectangle equals 1 : 2.236 ... . Its reciprocal ($\sqrt{5}/5$: 1) equals 0.4472... : 1.

IX Root Rectangles Within a Square

A series of diminishing root rectangles can be generated from a quarter-arc that is drawn within a square.

- Draw a square (DBAC) of side 1. (Repeat figures 2 and 3.)

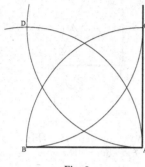

Fig. 2

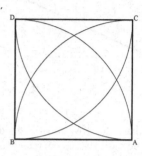

Fig. 3

- Locate the quarter-arc drawn from radius BA. Remove the other three.
- Draw the diagonal BC through the square DBAC.
- The quarter-arc (from radius BA) and the diagonal (BC) intersect at point E.
- Draw a horizontal line through point E that is parallel to line CD and intersects the line AC at point F and line DB at point G.
- Connect points G, B, A and F.

The result is a root-two rectangle (GBAF) with short and long sides of 1/√2 and 1 (fig. 26).

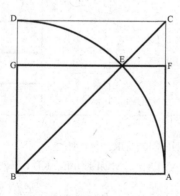

Fig. 26

GB:BA :: 1:√2

- Draw the diagonal BF through the 1 : √2 rectangle (GBAF).
- The quarter-arc (from radius BA) and the diagonal (BF) intersect at point H.
- Draw a horizontal line through point H that is parallel to line FG and intersects the line AF at point I and line GB at point J.

- Connect points J, B, A and I.

The result is a root-three rectangle (JBAI) with short and long sides of $1/\sqrt{3}$ and 1 (fig. 27).

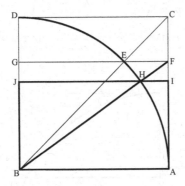

JB:BA :: $1:\sqrt{3}$

Fig. 27

- Draw the diagonal BI through the $1:\sqrt{3}$ rectangle (JBAI).
- The quarter-arc (from radius BA) and the diagonal (BI) intersect at point K.
- Draw a horizontal line through point K that is parallel to line IJ and intersects the line AI at point L and line JB at point M.
- Connect points M, B, A and L.

The result is a root-four rectangle (MBAL) with short and long sides of $1/\sqrt{4}$ and 1 (fig. 28).

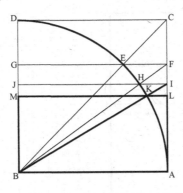

MB:BA :: $1:\sqrt{4}$

Fig. 28

- Draw the diagonal BL through the 1 : √4 rectangle (MBAL).
- The quarter-arc (from radius BA) and the diagonal (BL) intersect at point N.
- Draw a horizontal line through point N that is parallel to line LM and intersects the line AL at point O and line MB at point P.
- Connect points P, B, A and O.

The result is a root-five rectangle (PBAO) with short and long sides of 1/√5 and 1 (fig. 29).

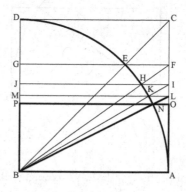

PB:BA :: 1:√5

Fig. 29

The long side of each root rectangle is 1. The short sides progress in the series 1/√2, 1/√3, 1/√4, 1/√5... (fig. 30).

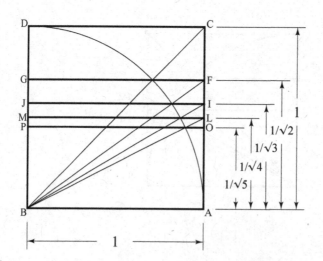

DB:BA ::1: 1

GB:BA ::1:√2

JB : BA ::1:√3

MB:BA ::1:√4

PB:BA :: 1:√5

Fig. 30

X Complementary Areas

In Hambidge's system, each rectangle has a reciprocal and each rectangle and reciprocal have complementary areas. These component shapes become evident when the rectangle is produced within a unit square.

Definition:

The **complementary area** is the area that remains when a rectangle is produced within a unit square. If the rectangle exhibits properties of dynamic symmetry, its complement will also.[14]

XI Complementary Patterns in the Root-Two Rectangle

- Draw a square (DBAC) of side 1. (Repeat figures 2 and 3.)

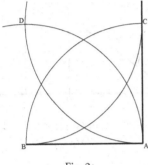

Fig. 2

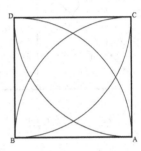

Fig. 3

- Locate the root-two rectangle GBAF of sides $1/\sqrt{2}$ and 1, as shown. (Repeat figure 26.)

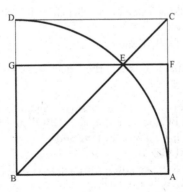

GB:BA :: $1/\sqrt{2}$:1

Fig. 26

- Locate the rectangle DGFC that remains.

Rectangle DGFC is the complement of rectangle GBAF.

- Locate point E where the diagonal BC intersects the top edge of the root-two rectangle (GBAF).
- Draw a vertical line through point E that is perpendicular to line FG and intersects the line BA at point Q and line CD at point R.
- Locate the square GBQE within the rectangle (GBAF) and the square REFC within the complementary area DGFC (fig. 31).

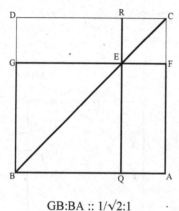

GB:BA :: 1/√2:1

Fig. 31

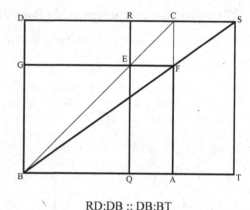

RD:DB :: DB:BT
1/√2:1 :: 1:√2

Fig. 32

- Draw the diagonal BF through the root-two rectangle (GBAF).
- Extend the diagonal from point F until it intersects the extension of line DC at point S.
- From point S, draw a line perpendicular to line SD that intersects the extension of line BA at point T.
- Connect points D, B, T and S.

The rectangle (DBTS) that results is in the ratio 1 : √2 (fig. 32).

DBAC is a unit square.
Rectangle DBTS is a root-two rectangle of sides 1 and √2.
Point E on the diagonal BC locates the squares GBQE and REFC.
Rectangle GBAF is a reciprocal of sides 1/√2 and 1.
Rectangles GBAF and RDBQ are equal
Rectangle DGFC is the complement of the reciprocal (GBAF).
Rectangles DGFC and QACR are equal.
Rectangles DGER and QAFE are equal.
Rectangles QAFE and ATSC share common diagonals and are similar.
The root-two rectangle DBTS divides into two reciprocals (RDBQ and SRQT).

Definition:

Rectangles are **similar** if their corresponding angles are equal and their corresponding sides are in proportion. Similar rectangles share common diagonals.

XII Complementary Patterns in the Root-Three Rectangle

- Draw a square (DBAC) of side 1. (Repeat figures 2 and 3.)

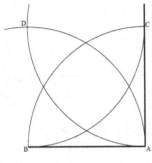

Fig. 2

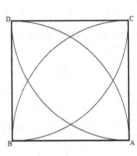

Fig. 3

- Locate the root-three rectangle JBAI of sides $1/\sqrt{3}$ and 1, as shown. (Repeat figures 26 and 27.)

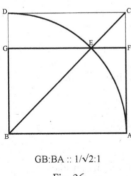

GB:BA :: $1/\sqrt{2}$:1

Fig. 26

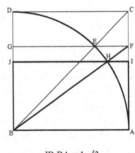

JB:BA :: $1:\sqrt{3}$

Fig. 27

- Locate the rectangle DJIC that remains.

Rectangle DJIC is the complement of rectangle JBAI.

- Locate point U where the diagonal BC intersects the top edge of the root-three rectangle (JBAI).

- Draw a vertical line through point U that is perpendicular to line IJ and intersects the line BA at point V and line CD at point W.
- Locate the square JBVU within the rectangle (JBAI) and the square WUIC within the complementary area DJIC (fig. 33).

- Draw the diagonal BI through the root-three rectangle (JBAI).
- Extend the diagonal from point I until it intersects the extension of line DC at point X.
- From point X, draw a line perpendicular to line XD that intersects the extension of line BA at point Y.
- Connect points D, B, Y and X.

The rectangle (DBYX) that results is in the ratio 1 : √3 (fig. 34).

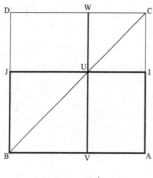

JB:BA :: 1/√3:1

Fig. 33

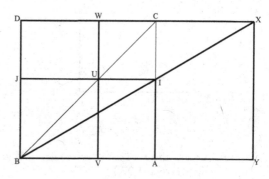

WD:DB :: DB:BY
1/√3:1 :: 1:√3

Fig. 34

DBAC is a unit square.
Rectangle DBYX is a root-three rectangle of sides 1 and √3.
Point U on the diagonal BC locates the squares JBVU and WUIC.
Rectangle JBAI is a reciprocal of sides 1/√3 and 1.
Rectangles JBAI and WDBV are equal
Rectangle DJIC is the complement of the reciprocal (JBAI).
Rectangles DJIC and VACW are equal.
Rectangles DJUW and VAIU are equal.
Rectangles VAIU and AYXC share common diagonals and are similar.
The root-three rectangle DBYX divides into three reciprocals.

XV Complementary Patterns in the Root-Four Rectangle

- Draw a square (DBAC) of side 1. (Repeat figures 2 and 3.)
- Locate the root-four rectangle MBAL of sides 1/√4 and 1, as shown. (Repeat figures 26-28.)

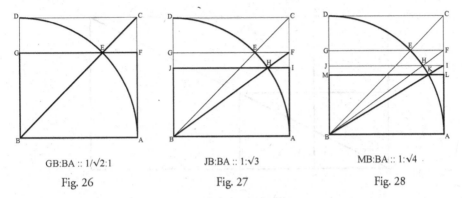

GB:BA :: 1/√2:1

Fig. 26

JB:BA :: 1:√3

Fig. 27

MB:BA :: 1:√4

Fig. 28

- Locate the rectangle DMLC that remains.

Rectangle DMLC is the complement of rectangle MBAL.

- Locate point Z where the diagonal BC intersects the top edge of the root-four rectangle (MBAL).
- Draw a vertical line through point Z that is perpendicular to line LM and intersects the line BA at point A′ and line CD at point B′.
- Locate the square MBA′Z within the rectangle (MBAL) and the square B′ZLC within the complementary area DMLC (fig. 35).

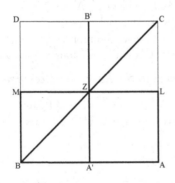

MB:BA :: 1/√4:1

Fig. 35

- Draw the diagonal BL through the root-four rectangle (MBAL).
- Extend the diagonal from point L until it intersects the extension of line DC at point C′.
- From point C′, draw a line perpendicular to line C′D that intersects the extension of line BA at point D′.
- Connect points D, B, D′ and C′.

The rectangle (DBD′C′) that results is in the ratio 1 : √4 (fig. 36).

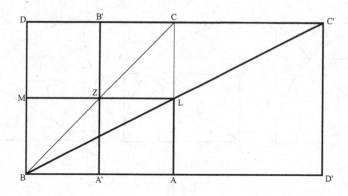

$$B'D:DB :: DB:BD'$$
$$1/\sqrt{4}:1 :: 1:\sqrt{4}$$

Fig. 36

DBAC is a unit square.
Rectangle DBD'C' is a root-four rectangle of sides 1 and √4.
Point Z on the diagonal BC locates the squares MBA'Z and B'ZLC.
Rectangle MBAL is a reciprocal of sides 1/√4 and 1.
Rectangles MBAL and B'DBA' are equal.
Rectangle DMLC is the complement of the reciprocal (MBAL).
Rectangles DMLC and A'ACB' are equal.
Rectangles DMZB' and A'ALZ are equal.
Rectangles A'ALZ and AD'C'C share common diagonals and are similar.
The root-four rectangle DBD'C' divides into four reciprocals.

XIV Complementary Patterns in the Root-Five Rectangle

- Draw a square (DBAC) of side 1. (Repeat figure 2 and 3.)
- Locate the root-five rectangle PBAO of sides 1/√5 and 1, as shown. (Repeat figures 26-29.)

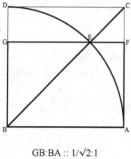

GB:BA :: 1/√2:1

Fig. 26

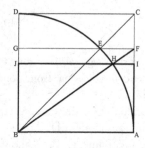

JB:BA :: 1:√3

Fig. 27

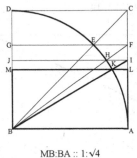

MB:BA :: 1:√4

Fig. 28

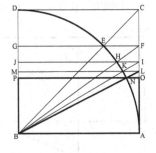

PB:BA :: 1:√5

Fig. 29

- Locate the rectangle DPOC that remains.

Rectangle DPOC is the complement of rectangle PBAO.

- Locate point E' where the diagonal BC intersects the top edge of the root-five rectangle (PBAO).
- Draw a vertical line through point E' that is perpendicular to line OP and intersects the line BA at point F' and line CD at point G'.
- Locate the square PBF'E' within the rectangle (PBAO) and the square G'E'OC within the complementary area DPOC (fig. 37).

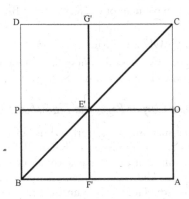

PB:BA :: 1/√5:1

Fig. 37

- Draw the diagonal BO through the root-five rectangle (PBAO).
- Extend the diagonal from point O until it intersects the extension of line DC at point H'.
- From point H', draw a line perpendicular to line H'D that intersects the extension of line BA at point I'.
- Connect points D, B, I' and H'.

The rectangle (DBI'H') that results is in the ratio 1:√5 (fig. 38).

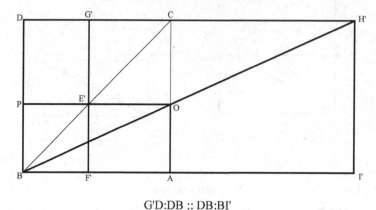

G'D:DB :: DB:BI'
1/√5:1 :: 1:√5

Fig. 38

DBAC is a unit square.
Rectangle DBI'H' is a root-five rectangle of sides 1 and √5.
Point E' on the diagonal BC locates the squares PBFE' and G'E'OC.
Rectangle PBAO is a reciprocal of sides 1/√5 and 1.
Rectangles PBAO and G'DBF' are equal.
Rectangle DPOC is the complement of the reciprocal (PBAO).
Rectangles DPOC and F'ACG' are equal.
Rectangles DPE'G' and F'AOE' are equal.
Rectangles F'AOE' and AI'H'C share common diagonals and are similar.
The root-five rectangle DBI'H' divides into five reciprocals.

XV: Application: Matthew Carpet Page, the Lindisfarne Gospels

One of the great painted manuscripts of early Christian Britain is the illuminated book known as the Lindisfarne Gospels. Produced in honor of Saint Cuthbert between the late seventh and early eighth centuries at the monastic center of Lindisfarne on Holy Island,[15] it is believed to have been written and decorated essentially by a single hand, likely that of the Bishop Eadfrith. The Lindisfarne Gospels contain two versions of the four Gospel texts: the Latin text attributed to Eadfrith; and a translation in the mid-tenth century into Old English by the priest Aldred, inserted between the lines of original text. To some, the book is an *opus dei*, or work of God, the result of intensive labor, revelation and prayer.[16]

The manuscript is one of the outstanding works to survive the Insular period of the early Middle Ages (sixth to ninth centuries) in Ireland and Britain, an era noted for book painting, jewelry, building design and sculpture and characterized by rich ornamentation and the blending of Celtic, Germanic, early Christian, Mediterranean and other influences. Insular manuscript art typically integrates script, text and decoration. Detailed patterns combine zoomorphic figures, plaited interlace, knot and spiral work [Brown 1994, 74, Brown 2003, 272].

The Lindisfarne Gospels contain fifteen primary decorated pages, including five decorative cross-carpet pages, so named for their similarity to oriental carpets, in which a cross is featured as the dominant motif, against a background of interlace fill. The manuscript contains 259 leaves or folios of fine calfskin vellum, assembled in groups of eight and one additional flyleaf. Back-pages reveal traces of prickings, compass holes, divider marks, rulings and grid lines, suggesting geometric guides to layout and design composition. Michelle Brown, former Curator of Western Manuscripts at The British Library, believes that preliminary designs were drawn in reverse on the back and viewed with the aid of backlighting. Back-drawings exist for almost every decorative element in the manuscript, permitting guidelines to be viewed even after pigment is applied to the front.[17]

The Matthew carpet page (folio 26v) introduces the Gospel of Matthew and is the longest of the illuminated carpet pages, measuring 186 mm x 249 mm across a double rectilinear frame [Bruce-Mitford in *Evangelium Quattuor Codex Lindisfarnensis* 1960, 2: 232]. Six terminals are laid in the form of a Latin cross, in 4 x 3 fashion. Five terminals resemble chalices or bells; the sixth terminal at the hub is circular. All have centers marked by circles of blank vellum, possibly intended for gilding. The cross and background panels consist of bird, animal and abstract interlace and spiral work.[18] It is noteworthy that the distance between adjacent terminal centers is a uniform 59 mm, except the lowest terminal, which measures 64 mm from the one directly above [Stevick 1994, 143, Brown 2003, 298, 325-327]. (See figure 39a, below, and Fig. 39b.)

Fig. 39a. The Matthew Cross-Carpet

At least three scholars have proposed specific geometric frameworks for the layout of the Matthew carpet page.[19] Rupert L. S. Bruce-Mitford, whose commentary accompanies a 1956 facsimile of the manuscript, utilizes prickings and other back page markings to establish a grid of circles and rulings that situate specific ornamental motifs. He does not delineate a primary spacing unit, nor does he resolve the proportions of the overall frame, the location and relative sizes of key elements, or the "irregular" placement of the lowest medallion [Bruce-Mitford in *Evangelium Quattuor Codex Lindisfarnensis* 1960, 2: 226-30, fig. 59, Guilmain 1987, 41].

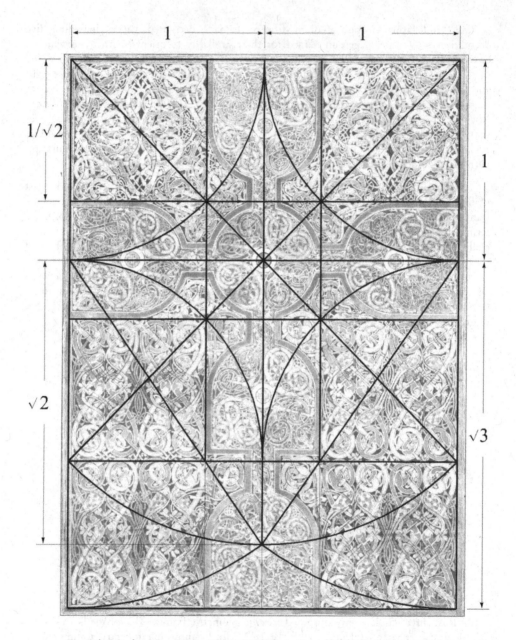

Fig. 39b. Matthew Cross-Carpet Page. Lindisfarne Gospels, fol. 26v, with geometric overlay by Rachel Fletcher

The scheme proposed by Jacques Guilmain identifies a primary spacing unit that is used throughout. An overall grid of 61 x 82 units encompasses the double frame and situates individual panels.[20] Guilmain's intent is to illuminate the design's additive and repetitive components, where patterns and combinations of patterns repeat within a defined space. His proposal is consistent with regularly spaced pricking and rulings that are evident on the blank recto of the folio [Guilmain 1987, 41-47].

Of the three, Robert D. Stevick alone proposes how individual elements relate by geometric proportion to the composition as a whole. His analysis is supported by traces on the reverse of 45° diagonal rulings that intersect at all six terminal centers. Stevick observes that the outside measure of the double frame (249 x 186 mm) is slightly more than 4 : 3 in ratio and proposes a 3 : √5 rectangle for its proportions [Stevick 1994, 142- 150].

Stevick's diagonal grid extends slightly beyond the inside frame on three sides, but the baseline of the composition is unresolved. And while the grid locates all six terminal centers, the relative sizes and proportions of masses are not addressed. The geometric layout in figure 39b proposes an additional layer of interpretation that complements Stevick's and other studies, describing the proportions of individual components in relation to one another and to the composition as a whole. A striking interplay of elements of √2 and √3 dynamic symmetry is evident.[21]

The study is confined to the composition within the double frame of the Matthew carpet page.

- Draw a square whose top edge coincides with the top inside edge of the double frame.
- Draw vertical and horizontal lines through the center of the square. Extend the vertical center line indefinitely.
- Draw 45° diagonal lines through the center of the square.

The center of the square locates the terminal center at the hub of the Matthew cross.

- Place the compass point at any corner of the square and draw a quarter-arc of radius equal to half the length of the square, as shown.
- Repeat at the remaining three corners.
- Locate the four points where the quarter-arcs and diagonals intersect.
- Draw vertical and horizontal lines through all four points.

The vertical and horizontal lines divide the Matthew carpet page into respective cross and background panels.[22]

- Place the compass point at the center of the square. Open the compass to a bottom corner of the square and draw a quarter-arc, as shown.

The quarter-arc and vertical center line intersect at the center of the lowest terminal of the Matthew cross.

- Place the compass point at the midpoint of the left edge of the square. Open the compass to the center of the quarter-arc, as shown. Draw an arc that intersects the extension of the left edge of the square.
- Repeat from the right edge of the square.

- Connect the two points of intersection.

The baseline that results locates the bottom inside edge of the double frame of the Matthew carpet page.

Elements of √2 and √3 dynamic symmetry appear to dominate the composition. Root-two rectangles appear on either side of the vertical center line, marked above by the center of the Matthew cross, and below by the center of the cross's lowest terminal. Reciprocal root-two rectangles are contained within each quarter square. The remaining complementary areas define the width of the Matthew cross. The total composition divides at the center of the cross into a square, above, and a root-three rectangle, below, on either side of the vertical center line. The study is supported on the blank recto of the folio by prickings and rulings that coincide with key elements of the geometric layout.[23]

It is not self-evident that the artist of the Matthew carpet page adopted any of the proposed geometric techniques, for the ability to demonstrate how a design can be constructed does not prove that it was done in that fashion.[24] But the precision with which dynamic √2 and √3 proportions appear in the composition is compelling. It is interesting to note that the Latin cross features six distinct terminals, as if to suggest the faces of an unfolded cube. The fact that √2 and √3 proportions are also present in the cube implies a new layer of symbolism. (See fig. 14.) Stevick observes that the leaves of the illuminated manuscript are in 1 : √2 ratio [Stevick 1994, 90]. Designed to lie flat, they form a root-two rectangle divided into two reciprocals.

XVI Summary

When a square divides into diminishing root rectangles, or when root rectangles expand from a square, elements of dynamic symmetry become apparent. We have experienced the relative spatial characteristics of root-two, -three, -four, and -five systems of proportion and are familiar with diagonals, reciprocals, complementary areas and other components. We have observed these at play in the Matthew carpet page of the Lindisfarne Gospels. In future columns, we explore new ways to utilize these elements in dynamic space plans.

Notes

1. P. H. Scholfield's name for "dynamic symmetry" is "repetition of similar figures," which he traces to the works of Vitruvius and Alberti, and in the late eighteenth and nineteenth centuries to the theories of A. Barca, William Watkyss Lloyd and August Thiersch. For Thiersch, "the repetition of the fundamental form of the plan through its subdivisions" contributes to the best of art throughout history [quoted in Scholfield 1958, 102]. Scholfield identifies such other theorists and practitioners as Heinrich Wölfflin, John Beverly Robinson, and more recently Harry Roberts, Sir Edwin Lutyens, Corbusier, William Schooling and Sir Theodore Cook. See [Scholfield 1958, 102-120].
2. [Hambidge 1967, xii, xiv-xvi.] In contrast, static symmetry is accomplished by dividing a linear measure into even multiple units or by the radial subdivision of regular geometric figures, crystals or flowers. Dynamic symmetry, found in shell growth or the arrangement of leaves on a plant, is a more vital and flexible organizing system [Hambidge 1960, 3, Scholfield 1958, 117].

3. Hambidge was unfamiliar with ancient Greek when he chose the term "dynamic symmetry" to describe this characteristic. Later, he discovered Thomas Heath's edition of Euclid, where the Greek *dunamei summetros* or *summetrôs dunama*, previously translated "commensurable in power," is "commensurable in square" [Euclid 1956, III: 11 (bk. X, def. 2), Hambidge 1967, 129, Liddell 1940].

4. The long side of a root-four rectangle ($\sqrt{4}$ or 2.0) is commensurable in length, but the root-four rectangle exhibits dynamic symmetry in other ways.

5. The calculation of the diagonal BC is based on the Pythagorean Theorem. Triangle CDB is a right triangle. $CD^2 + DB^2 = BC^2$ [$1^2 + 1^2 = 2$]. Thus, BC = $\sqrt{2}$. The diagonal of a square of side 1 is equal to $\sqrt{2}$. The diagonals that appear in figures 6, 8, 10 and 12 can be calculated in similar fashion. See [Fletcher 2005b, 44-45] for more on the Pythagorean Theorem.

6. [Berriman 1969, 29.] John Michell proposes a more comprehensive system, but specifies different values for each unit of measure. In Michell's version, one Egyptian remen (the side length of the square) equals 1.2165 feet or 14.598 inches. One royal cubit (the long length of the root-two rectangle) equals 1.72 feet or 20.64 inches. One Palestinian cubit (the long length of the root-three rectangle) equals 2.107 feet or 25.284 inches. One Roman pace (the long length of the root-four rectangle) equals 2.433 feet or 29.196 inches. One megalithic yard (the long length of the root-five rectangle) equals 2.72 feet or 32.64 inches. Virtually one yard (the long length of the root-six rectangle) equals 2.98 feet or 35.76 inches [Michell 1972, 106]. For more on the Megalithic yard, discovered by Alexander Thom, see [Thom 1976, 34].

7. In fig. 14, the three lengths of 1, $\sqrt{2}$ and $\sqrt{3}$ form a right triangle. In fig. 15, the three lengths of 1, $\sqrt{4}$ and $\sqrt{5}$ form a right triangle. In fig. 16, the three lengths of 1, $\sqrt{5}$ and $\sqrt{6}$ form a right triangle. This is evident by the Pythagorean Theorem.

8. By the Pythagorean Theorum, BF equals $\sqrt{3}$ and EG equals $\sqrt{3}/\sqrt{2}$.

9. The radius vector is the variable line segment drawn to a curve or spiral from a fixed point of origin (the pole or eye) [Simpson 1989]; see also [Fletcher 2004, 105]. For more on the root-two rectangle and its inherent proportions, see [Fletcher 2005b, 55-56].

10. By the Pythagorean Theorum, BH equals $\sqrt{4}$ and GI equals $\sqrt{4}/\sqrt{3}$.

11. For more on the root-three rectangle and its inherent proportions, see [Fletcher 2004, 102-105; and 2005a, 153-157].

12. By the Pythagorean Theorem, BJ equals $\sqrt{5}$ and IK equals $\sqrt{5}/\sqrt{4}$.

13. By the Pythagorean Theorem, BL equals $\sqrt{6}$ and KM equals $\sqrt{6}/\sqrt{5}$.

14. Hambidge says, "[The] complement of a shape is that area which represents the difference between the selected figure and unity. The complement of any number is obtained by dividing the number into unity to find the reciprocal and subtracting that reciprocal from unity" [1967, 128].

15. Holy Island is located in northeastern England, Northumbria.

16. [Backhouse 1981, 7, 12, 14, Brown 2003, 280, 283-4, 397, 399.] Janet Backhouse dates the manuscript just prior to 689 when Eadfrith became bishop of Lindisfarne. Michelle Brown's more recent investigation dates Eadfrith's work to 710-720. The book resides today at The British Library in London, where it is known as Cotton MS Nero D.iv.

17. [Brown 2003, 200-203, 217-220, 290-98.] One exception is the series of Canon Tables that repeat a basic layout and reveal prickings and rulings in another fashion.

18. Eric George Millar [1923, 37] counts sixty-six animals and fifty-eight birds in all. For Brown on the theological significance of the Matthew and other carpet pages, see [Brown 2003, 297-298, 312-327].

19. Others such as John Romilly Allen, George Bain and Eric George Millar [1923, 36-38] offer guides or techniques for accomplishing ornamental elements, but do not consider individual elements as they relate to the composition as a whole. See [Guilmain 1987, 21-22, Stevick 1994, 146]. In addition, Swenson [1978, 9-11, 14-15, fig. 12] provides examples of rotational and reflexive symmetry.

20. In Guilmain's scheme, a primary unit measuring 3.04 mm wide and 3.05 mm long is required to conform to Bruce-Mitford's measure of 186 x 249 mm for the double frame.

21. The study utilizes scaled photographs of the original manuscript, courtesy of Janet Backhouse, The British Library. COTTON.NERO.D.IV f26v © British Library Board. All Rights Reserved. Geometric overlay: Rachel Fletcher.

22. The baseline of the square does not align with the base of its adjacent terminal, but is indicated on the reverse by a distinct ruled line and prickings.

23. Rulings that do not delineate a finished pattern likely serve to advance a geometric layout. These include a full-length, ruled vertical center line and a full-length ruled horizontal line that coincides with the baseline of the square. Traces of horizontal rulings through the center and left terminals of the Matthew cross coincide with the horizontal line through the center of the square. Compass holes or prickings are evident at the midpoint of the left edge of the square and at the intersection of the left edge and baseline. Possibly, these were used to draw a √3 radius arc. A ruled horizontal line coincides with the bottom inside edge of the inner frame. Numerous prickings suggest its importance to the layout.

24. [Guilmain 1987, 22-23, Stevick 1994, 145-146]. Brown concedes that technical innovations used to produce the Lindisfarne Gospels, such as leadpoint, backlighting and reversed design, developed under solitary working conditions and may not have been widespread. Leadpoint, the precursor to pencil, enabled the design to be laid out *in situ*, rather than through the cumbersome transfer from secondary vehicles, such as slate, stone, wood or wax tablets used by other artists of the period [Brown 224-226, 290, 298].

References

BACKHOUSE, Janet. 1981. *The Lindisfarne Gospels*. Oxford: Phaidon Press.

BERRIMAN, A. E. 1969. *Historical Metrology*. 1953. Reprint. New York: Greenwood Press.

BROWN, Michelle P. 1994. *Understanding Illuminated Manuscripts: A Guide to Technical Terms*. Los Angeles: Getty Publications and The British Library.

———. 2003. *The Lindisfarne Gospels: Society, Spirituality and the Scribe*. Toronto: University of Toronto Press.

EUCLID. 1956. *The Thirteen Books of Euclid's Elements*. Thomas L. Heath, ed. and trans. Vols. I-III. New York: Dover.

Evangelium Quattuor Codex Lindisfarnensis. 1956, 1960. 2 vols. (1) facsimile; (2) commentary by T. D. Kendrick, T. J. Brown, R. L. S. Bruce-Mitford, et al. Olten and Lausanne: Urs Graf.

FLETCHER, Rachel. 2004. Musings on the Vesica Piscis. *Nexus Network Journal* 6, 2 (Autumn 2004): 95-110.

———. 2005a. Six + One. *Nexus Network Journal* 7, 1 (Spring 2005): 141-160.

———. 2005b. The Square. *Nexus Network Journal* 7, 2 (Autumn 2005): 35-70.

GUILMAIN, Jacques. 1987. The Geometry of the Cross-Carpet Pages in the Lindisfarne Gospels. *Speculum* 62, 1 (January 1987): 21-52.

HAMBIDGE, Jay. 1960. *Practical Applications of Dynamic Symmetry*. 1932. New York: Devin-Adair.

———. 1967. *The Elements of Dynamic Symmetry*. 1926. New York: Dover.

LIDDELL, Henry George and Robert SCOTT, eds. 1940. *A Greek-English Lexicon*. Henry Stuart Jones, rev. Oxford: Clarendon Press. Perseus Digital Library Project. Gregory R. Crane, ed. Medford, MA: Tufts University. 2005. http://www.perseus.tufts.edu

MICHELL, John. 1972. *City of Revelation: On the Proportions and Symbolic Numbers of the Cosmic Temple*. New York: David McKay Company.

MILLAR, Eric George. 1923. *The Lindisfarne Gospels*. London: British Museum.

SCHOLFIELD, P. H. 1958. *The Theory of Proportion in Architecture*. Cambridge: Cambridge University Press.

SIMPSON, John and Edmund Weiner, eds. 1989. *The Oxford English Dictionary*. 2nd ed. OED Online. Oxford: Oxford University Press. 2004. http://www.oed.com/

STEVICK, Robert D. 1986. Crosses in the Lindisfarne and Lichfield Gospels. *Gesta* **25**, 2 (1986): 171-184.

———. 1994. *The Earliest Irish and English Bookarts: Visual and Poetic Forms Before A.D. 1000.* Philadelphia: University of Pennsylvania Press.

SWENSON, Inga Christine. 1978. The Symmetry Potentials of the Ornamental Pages of the Lindisfarne Gospels. *Gesta* **17**, 1 (1978): 9-17.

THOM, A. 1976. *Megalithic Sites in Britain.* 1967. Reprint. Oxford: Oxford University Press.

About the geometer

Rachel Fletcher is a theatre designer and geometer living in Massachusetts, with degrees from Hofstra University, SUNY Albany and Humboldt State University. She is the creator/curator of two museum exhibits on geometry, "Infinite Measure" and "Design By Nature". She is the co-curator of the exhibit "Harmony by Design: The Golden Mean" and author of its exhibition catalog. In conjunction with these exhibits, which have traveled to Chicago, Washington, and New York, she teaches geometry and proportion to design practitioners. She is an adjunct professor at the New York School of Interior Design. Her essays have appeared in numerous books and journals, including *Design Spirit*, *Parabola*, and *The Power of Place*. She is the founding director of Housatonic River Walk in Great Barrington, Massachusetts, and is currently directing the creation of an African American Heritage Trail in the Upper Housatonic Valley of Connecticut and Massachusetts.

Sarah Maor

Igor M. Verner

Department of Education
in Technology & Science
Technion – Israel Institute
of Technology
Haifa, 32000 ISRAEL
ttrigor@tx.technion.ac.il

Keywords: architecture
education, mathematics,
design studio, geometrical
forms, project based
learning

Didactics

Mathematical Aspects in an Architectural Design Course: The Concept, Design Assignments, and Follow-up

Abstract. This paper considers a Mathematical Aspects in Architectural Design course in a college of architecture. The course is based on experiential learning activities in the design studio. It focuses on designing architectural objects, when the design process is tackled from three geometrical complexity directions: tessellations, curve surfaces, and solids intersections. The students perform seminars, exercises, and projects in which they analyse and develop geometrical forms and implement them in design solutions. Students achievements in design and mathematics are assessed. The course follow-up indicated that the students used mathematics as a source of complex geometrical forms and a tool for designing efficient solutions.

Introduction

Architecture education, from the pedagogical perspective, is grounded on the methodology of constructivism considering learning as an active process in which the learner constructs knowledge through practice and interaction with the environment. Teymer [1996] explicitly articulates these roots, stating the need to educate architecture students toward self directed, holistic, profound, and reflective reasoning of the environment.

When discussing the architecture curriculum, Banerjee and De Graaf [1996] point out that it consists of two main blocks, namely, the block of preparatory disciplines and the block of problem oriented integrated disciplines. The preparatory disciplines are traditionally taught with frontal lessons, while the integrated disciplines are given through the project based learning in the architectural design studio. The authors focus on the preparatory structures course and discuss students' difficulties in understanding concepts of statics. These difficulties cause lack of ability to apply the concepts in their studio projects.

Mathematics is one of the preparatory disciplines and is traditionally delivered through frontal teaching. Architecture educators [Salingaros 1999] notice students' difficulties in understanding mathematics concepts and in solving mathematical problems of structures design, and there is a call for change.

Our longitudinal study examines the way to close the gap between the two blocks of disciplines and overcome the disharmony between learning in frontal classrooms and in design studios. Our previous papers [Verner and Maor 2003; Verner and Maor 2006] discuss two main approaches to teaching mathematics with applications in various contexts: Realistic Mathematics Education (RME) and Mathematics as a Service Subject (MSS). From the epistemological view both approaches are grounded on the methodology of constructivism.

By involvement in solving mathematical problems related to architecture structures, and in geometrical design projects, the students gradually build their mathematical knowledge and develop ability to use it in architecture design. Looking metaphorically, the role of teaching in supporting this constructivist learning is similar to the function of scaffolding as a temporary framework providing stability and efficiency during the building construction.

The RME approach is implemented in our introductory calculus course [Verner and Maor 2003] in which learning mathematical concepts is supported by solving applied problems relevant to architecture studies. Our second, more advanced course Mathematical Aspects in Architectural Design (MAAD) implements the MSS approach [Verner and Maor 2006] when the focus is on developing the ability to apply mathematical methods in performing architecture design assignments. In this paper we summarize the six-year experience of teaching the introductory course and the three-year practice of the MAAD course. During this period the college has become an academic institution educating students for a Bachelor of Science degree in architecture and the two courses continue to be a part of the curriculum.

Curriculum development principles

When developing the MAAD course curriculum we followed the planning and evaluation model for a project-based curriculum in architecture proposed by Teymur [1992]. Accordingly, the didactical principles, practice, outcomes, and evaluation are developed through the following design activities:

- A. Define the concept (answering the question – Why?)
- B. Design projects assignments (answering the question – What?)
- C. Develop a learning environment (answering the questions – How and Where?)
- D. Plan the course framework and management way (answering the questions – How and When?).

A. When defining the concept of our course, the main question was to find a way of observing different complex geometrical forms in a systematic way. Our approach [Verner and Maor 2006] is to consider the three directions of geometrical complexity in architectural objects: tessellations, curved surfaces, solids intersections (fig. 1).

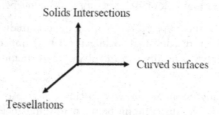

Fig. 1. The three directions of geometrical complexity

We believe that performing geometrical design projects in each of these directions facilitates students' ability and motivation to integrate complex geometrical solutions in their architecture design projects. At the next step mathematical concepts relevant to geometrical design for different directions of geometrical complexity were selected:

- Arranging regular shapes to cover the plane (tessellations).
  Related mathematical concepts: proportions, symmetry, harmonic dimensions, golden section, Fibonacchi numbers, logarithmic spirals, polygons, modularity, curve smoothness, shape displacements, rotations, reflections, and combinations.
- Shaping curve lines and surfaces.
  Related mathematical concepts: folded plates, barrel vaults, domes, and shells.
- Analyzing solids' intersections.
  Related mathematical concepts: polyhedron, vertex, edge, envelop, facet, solids' composition and intersection.

B. To achieve the goals formulated at the first stage, the MAAD course includes exercises, seminars, and projects. They are partly discussed in [Verner and Maor, 2006] which presents the pilot course experience. The projects performed in the subsequent MAAD courses are presented below "Design Projects" section.

C. We found that the architectural design studio fits the requirements of the MAAD course because of the following reasons:

- The studio is familiar to students as an authentic environment for architectural design practices;
- The studio supports experiential learning and learning-by-doing processes [Schoen, 1988];
- The studio is suitable for experimentation with materials and physical modelling;
- The students in the studio learn theoretical concepts which they need to apply in their design practice.

D. The methods of teaching mathematics and geometrical design were integrated in the way presented in Table 1.

Didactical methods	Instructional objectives
Student seminars in geometrical analysis of structures, guided by the teacher	1. Acquiring mathematical concepts and linking them to architecture design concepts 2. Understanding the connection between architecture design and technology 3. Formal defining of mathematical concepts 4. Identifying mathematical concepts in architectural objects
Geometrical problem solving	1. Acquiring competences of applying geometrical concepts and methods 2. Geometrical analysis of physical models 3. Interpreting geometrical objects in the architecture context 4. Acquiring skills of building physical models
Project design and analysis	1. Training of divergent thinking through developing design alternatives. 2. Identifying, solving, and applying mathematical problems related to designing the product. 3. Gaining experience of creating and presenting the product
Peer and self-evaluation of the projects	1. Developing evaluation criteria 2. Evaluation and self-evaluation practice

Table 1. Didactical methods and instructional objectives

As follows from the table, the course applies four different didactical methods each of which has its instructional objectives. These instructional objectives were of three types: mathematical, architectural and integrated objectives. In the next section we present examples of the three design projects and the ways of achieving their instructional objectives.

Design Projects

Project 1. Arranging regular shapes to cover the plane (tessellations)

Tessellations are designed in architecture in order to cover floors, walls, roofs, and other architectural elements. They consist of flat geometrical objects formed by translation, rotation and intersection of basic figures. The coverings should be designed without overlapping and have minimal tailings. Frederickson [1997] pointed out two methods of dissecting geometrical figures into pieces and arranging tessellations: inscribing a figure in a certain tessellation module, and combining figures by joining vertices to compose a module. The additional geometrical concept central for these design was the proportion.

Fig. 2. The tessellation and its fragments designed by student A

Particular attention is paid to the golden section as a means to express harmony and aesthetics from ancient Greek architecture to Le Corbusier's Modulor [Le Corbusier 1968; Huylebrouck and Labarque 2002]. Ranucci [1974] studied these mathematical ideas and procedures of tessellation design implemented in Escher's artworks.

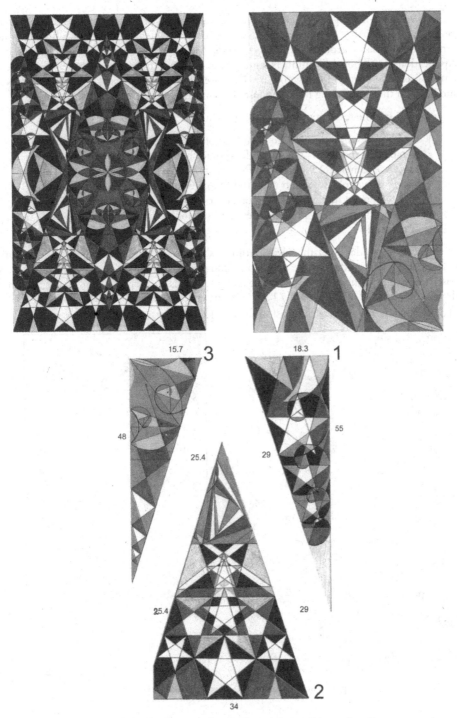

Fig. 3. The tessellation and its fragments designed by student B

The value of these geometrical concepts is recognized in the mathematics education. Boles and Newman [1990] developed a curriculum which studied plain tessellations arranged by basic geometrical shapes with focus on proportions and symmetry. Applications of Fibonacci numbers and golden section in designing tessellations were emphasized. Following [Salingaros 2001], tessellation studies focus on designing one module and then replicating it to compose the covering.

In our course the tessellation project assignment was as follows:

> Design a tessellation of a floor surface of 34×55 m by means of identical rectangular modules. The module should be a periodic combination of various geometrical figures. Define proportions and dimensions of the figures using golden section ratio and Fibonacci numbers. Develop a concept of the designed module choosing one of the following metaphoric subjects: a temple, kindergarten, political message, harmony with nature, and musical impression.

Below we consider how the ideas outlined above were implemented in project work performed by four students. The tessellation designed by student A is shown in fig. 2. It is based on regular pentagons with five logarithmic spirals in each of them. The spirals are drawn by means of a sequence of 72°-72°-36° triangles related to the golden section [Boles and Newman, 1990, p. 186]: $2 \cdot \cos(36°) = \varphi = 1.618K$.

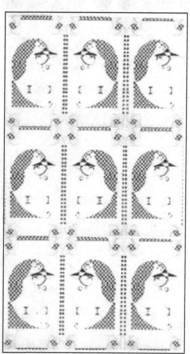

Fig. 4. The tessellations designed by students C and D

The module consists of these pentagons and the decagon created by joining ten pentagons. The decagon is dissected by the segments connecting the mid-points of its sides. Thus a smaller decagon is created and dissected in the same way, and so on creating also a Baravelle spiral [Boles and Newman 1990, 197]. Finally, the modules are combined composing the floor surface rectangle.

A design created by student B is presented in fig. 3. This student also used 72°-72°-36° triangles and the same procedure for drawing logarithmic spirals. But the original idea of this student was to dissect the triangles into triangle pieces (fig. 3, bottom) which constitute pentagons and pentagonal stars of different dimensions, composing the module (fig. 3, upper right), while the tessellation (fig. 3, upper left) is the combination of the modules.

Tessellations created by students C and D are shown in fig. 4. Student C created squares of two types. Squares of the first type are compositions of four straight line spirals [Boles and Newman 1990, 199], while squares of the second type have in the corners spirals, based on the Fibonacci rectangles. The tessellation of student D is based on images in which spirals are used to limit shading areas.

Project 2. Shaping curve lines and surfaces

Mathematical curved surfaces are used in architectural design as "the close link between form and structure, between geometry and the flow of forces in the structure" [Hanaor 1998, 147-148]. Curved surfaces that minimize deformation of structures under distributed loads are implemented in solutions existing in nature [Grosjean and Rassias 1992].

There is a long tradition of using mathematical surfaces in architecture. Gaudi systematically applied mechanical modeling to create geometrical forms and to examine their properties [Alsina and Gomes-Serrano 2002]. He also created 3D surfaces such as paraboloids, helicoids, and conoids by moving generator lines "in a dynamic manner" [Alsina, 2002, 89]. In architecture education, by studying mathematical surfaces the students are exposed to the construction of optimal structural elements.

In our course the second project focuses on using mathematical surfaces for roof design. The project assignment was described in our previous paper [Verner and Maor 2006] which presented results of the course given for the first time. Since then, the course has become part of the academic curriculum. Here we present some advanced ideas implemented in students' projects. Some of these ideas the students found in modern architecture such as Buckminster Fuller's Geodesic Dome, Tully Daniel's saddles (Hypar surfaces), and Santiago Calatrava's geometrical surfaces.

For the reader's convenience we repeat the project assignment definition from our previous paper:

> Design a plan of a gas station. Start from a zero level plan including access roads, parking, pumps, car wash, coffee shop, and an office. Design a top covering for the pumps area, or the roof of the coffee shop and office building. Find a design solution answering the stability, constructive efficiency, complexity, and aesthetics criteria.

Four examples of project works performed by students in the Spring 2006 course are presented in fig. 5. These examples implement different variations of mathematical surfaces of two types: Sarger and Conoid segments. Thee segments are described by the following equations:

$$Z = y\frac{f}{L}\left(1-\frac{4x^2}{l^2}\right) \qquad Z = d\left(1-\frac{y^2}{L^2}\right)+f\cdot\frac{y^2}{L^2}\cos\left(\frac{\pi\cdot L\cdot x}{l\cdot y}\right)$$

(Conoid segment)   (Sarger segment)

In these equations parameters l and f are the width and height of the generator line, L is the segment length, d is the end-point height.

The physical models were constructed by the students after precise calculation. Student E composed her solution from three Sarger segments combined in the form of flower petals. The roof created by student F consists of six Conoid segments, three at each side. Student G also uses six Conoids but combines them in a "wavy" form. Student H creates the roof from two Conoid segments of different dimensions.

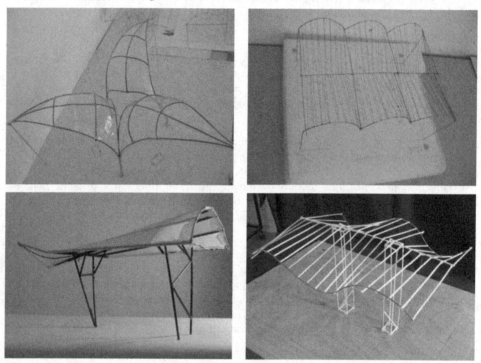

Fig. 5. The roof physical models designed by four students (E – H)

Project 3. Analyzing the intersections of solids

Many architectural buildings combine solid elements of different forms answering diverse functional needs. Combining these elements requires solutions of solid intersection problems.

Alsina [2002, 119-126] considered the design of complex three-dimensional forms by intersecting various geometrical forms. He analyzed the use of these forms in Gaudi's creations in order to achieve functional purposes such as light effects or symbolic expressions. Burt [1996] examined integrating and subdividing space by different types of polyhedral elements. He emphasized that this design method can provide efficient architectural solutions.

In architecture education solids and their intersections are studied in the morphology course [Haspelmath 2002]. Traditionally, this course does not include analytic calculations of intersections of solids required for precise design. Our course addresses this issue by offering the third project assignment which is formulated as follows:

> Select a known public building which was designed with solids intersections. The project requirements are:
>
> — Seminar on the building design process;
> — Mathematical analysis of the solids intersection;
> — Building a model which accurately presents solids intersection in the building;
> — Designing an additional functional module in the solids intersection area.

A sample project work performed by one of the students is presented in fig. 6. When selecting a public building (a museum in the north of Israel) the student recognized the solids intersection part of the structure (cylinder-pyramid). She measured building's dimensions and extracted geometrical data from the architecture design plans. After mathematical analysis of these data (fig. 6a) the student constructed the precise physical model of the building (fig. 6b).

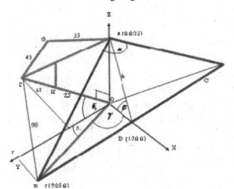

Fig. 6. A building composed of different solid elements:
a) the geometrical scheme; b) the physical model

Project Assessment

The students in the course perform the three projects and report them in project portfolios. Students' achievements in design and in mathematics are assessed using different evaluation criteria which fit specific characteristics of the projects. Our previous paper presents evaluation results only for the second project (curved surfaces). In this paper we present and compare evaluation results for the three course projects. The achievements in mathematics are assessed by the course lecturer (Maor), while the design achievements are assessed by an architect teaching the Design Studio course.

Tables 2A and 2B present mean grades in the three course projects, for design and mathematics criteria mentioned in the tables.

Project 1	Subject expression	Conception & application	Geometrical variety	Graphic representation	Design grade
Mean	83.1	89.0	77.4	64.6	74.7

Project 2	Efficiency	Aesthetics	Functionality	Program quality	Design grade
Mean	76.9	85.6	73.5	87.5	78.9

Project 3	Structure selection	Geometrical form	Model aesthetics	Additional module	Design grade
Mean	90.0	68.0	76.3	77.5	77.0

Table 2A. Design assessment grades

Project 1	Problems perception	Calculus application	Modularity application	Proportion harmony	Calculations	Drawings precision	Parametric solutions	Math grade
Mean	72.1	56.2	78.4	87.5	80.2	79.4	58.1	73.1

Project 2	Dimensions calculation	Surface parameters	Roof calculation	Building a model	Model precision	Model analysis	Geometrical complexity	Math grade
Mean	76.9	68.8	81.3	76.3	81.9	81.9	85.0	78.9

Project 3	Dimensions calculation	Use of parameters	Intersection calculation	Building a model	Model precision	Intersection analysis	Geometrical complexity	Math grade
Mean	73.8	68.8	77.5	75.0	80.0	77.5	82.5	76.4

Table 2B. Mathematics assessment grades

Tables 2A and 2B reveal the following features:

1. The mean grades of the three projects are similar in design (74.7-78.9) and in mathematics (73.1-78.9). The project 1 grades were slightly lower than of projects 2 and 3. A possible reason is that in the project the students dealt with tessellations design for the first time.
2. Mean grades for the use of parameters in the three projects were lower than for other mathematics criteria. These difficulties originated from the students' mathematical background.
3. Close correlation between the individual design and mathematics grades was found in project 1 ($\rho = 0.665$) and in project 2 ($\rho = 0.698$). This result indicates the tight integration of design and mathematical aspects of these course projects. In project 3 the correlation between the grades was lower ($\rho = 0.398$). A possible explanation is that project 3 does not include a design component, but focuses on analyzing existing structures.

Learning Activities in the MAAD Projects

An educational study was conducted in conjunction with the MAAD course development. Its methodology is discussed in our previous paper [Verner and Maor 2006] and presented in detail in the doctoral dissertation [Maor 2005]. One of the main results of this study is characterizing learning activities in the MAAD course projects. The learning activities are characterized at each of the architectural design stages. A summary of mathematical characteristics specific for these learning activities is presented in Table 3.

Design stages	Specific characteristics of learning activities
1. Project idea	Understanding the design and mathematical requirements of the project assignment. Identifying mathematical concepts adequate for expressing the metaphoric, symbolic, and analogical aspects of the project idea.
2. Data presentation	Imagining geometrical forms suitable for representing the design concept. Experimentation with physical and simulative models of geometrical forms and their mathematical description.
3. Analysis of the data and the constraints	Analysis of dimensions, scales, constraints, and properties. Search for suitable geometrical forms.
4. Generation of design alternatives	Identifying mathematical criteria for evaluation of design alternatives. Iterative synthesis, evaluation, and revision of alternative geometrical solutions.
5. Definition of design criteria	Mathematical description of the aesthetics, proportion, efficiency, modularity, symbolism, and accuracy criteria.
6. Selection of the design solution	Mathematical analysis of alternatives, finding the optimal solution, and its substantiation.
7. Presentation of the solution	Drawing the designed architectural object based on calculating dimensions, scales, and functions. Building physical and simulative models. Writing the mathematics-in-design report.
8. Evaluation	Presenting the project to peers and defending the solution.

Table 3. Characteristics of learning at different design stages

Conclusion

The proposed Mathematical Aspects in Architectural Design (MAAD) course has been taught for three years and has become important part of the college architecture curriculum. The course offers mathematics studies grounded on the constructivistic methodology through learning by design activities in the design studio. This second-year course continues and relies on the first-year mathematics course, in which applied problems are integrated following the realistic mathematics education approach.

Addressing geometrical complexity of architectural objects from the three its different directions (tessellations, curved surfaces, and solids' intersections) provides students with the broad perspective of mathematics applications in designing architectural forms.

Students in the course learn to use mathematics as a source of creative solutions and as an instrument to answer design criteria, such as constructive efficiency, functionality, optimization, shape variety, stability, and preciseness.

Each of the three parts of the course includes a study of mathematical concepts and methods with connection to architecture, practice in solving mathematical problems, and a design project. The mathematical learning activities in the projects include: analytic description of metaphoric, symbolic, and analogical aspects of the project idea; search for suitable geometrical forms and their analysis; synthesis, evaluation, and revision of geometrical solutions; mathematical description of the criteria for aesthetics, proportions, efficiency, modularity, and accuracy; finding the optimal solution, and its substantiation; building physical and simulative models based on calculating dimensions, scales, and functions.

The projects in the course are assessed through analysis of activities, design solutions and mathematics applications. Design solutions are assessed following the existing practice of studio evaluation with regards to the following aspects: concept, planning/detailing, and representation/expression. Mathematics applications are assessed using the following criteria: perception of mathematical problems, solving applied problems, precision in drawing or building a physical model of geometrical objects, accuracy of calculations and parametric solutions. Results of the course assessment indicated that the students used various complex geometrical shapes as a source of creative and efficient design solutions of the three project assignments.

References

ALSINA, C. and GOMES-SERRANO J. 2002. Gaudian Geometry. Pp. 26-45 in *Gaudi. Exploring Form: Space, Geometry, Structure and Construction*, Daniel Giralt-Miracle, ed. Barcelona: Lunwerg Editores.

ALSINA, C. 2002. Conoids. Pp. 88-95 in *Gaudi. Exploring Form: Space, Geometry, Structure and Construction*, Daniel Giralt-Miracle, ed. Barcelona: Lunwerg Editores.

————. 2002. Geometrical Assemblies. Pp. 118-125 in *Gaudi. Exploring Form: Space, Geometry, Structure and Construction*, Daniel Giralt-Miracle, ed. Barcelona: Lunwerg Editores.

BANERJEE, H.K. and DE GRAAF, E. 1996. Problem-Based Learning in Architecture: Problems of integration of technical disciplines. *European Journal of Engineering Education* 21 (2): 185-196.

BOLES, M. and NEWMAN, R. 1990. *Universal Patterns. The Golden Relationship: Art, Math and Nature*. Massachusetts: Pythagorean Press.

BURT, M. 1996. *The Periodic Table of The Polyhedral Universe*. Haifa: Technion – Israel Institute of Technology.

FREDERICKSON, G. 1997. *Dissections: Plane and Fancy*. Cambridge: Cambridge University Press.

HANAOR, A. 1998. *Principles of Structures*. Oxford: Blackwell Science.

GROSJEAN, C.C. and RASSIAS, T.M. 1992. Joseph Plateau and his work. Pp. 3-17 in *The problem of Plateau*. NJ: River Edge.

HASPELMATH, M. 2002. *Understanding Morphology*. Oxford University Press: London.

HUYLEBROUCK, D. and LABARQUE, P. 2002. More True Applications of the Golden Number. *Nexus Network Journal* 4, 1: 45-58.

LE CORBUSIER. 1968. *Modulor*. Cambridge: MIT Press.

MAOR, S. 2005. Mathematical Aspects in Architectural Design in Training Practical Engineers. Doctoral Dissertation. Haifa: Technion – Israel Institute of Technology.

RANUCCI, E. R. 1974. Master of tessellations: M.C. Escher, 1898-1972. *Mathematics Teacher* 4: 299-306.

SALINGAROS, N.A. 1999. Architecture, patterns, and mathematics. *Nexus Network Journal* 1: 75-86.

SALINGAROS, N. A. and TEJADA, D. M. 2001. Modularity and the Number of Design Choices. *Nexus Network Journal* 3, 1: 99-109.

SCHOEN, D. 1988. The Architectural Studio as an Examplar of Education for Reflection-in-Action. *Journal of Architectural Education* 38, 1: 2-9.

TEYMUR, N. 1992. *Architectural Education: Issues in Educational Practice and Policy*. London: Question Press.

————. 1996. *City as Education*. London: Question Press.

VERNER, I and MAOR, S. 2003. The Effect of Integrating Design Problems on Learning Mathematics in an Architecture College. *Nexus Network Journal* 5, 2: 111-115.

————. 2006. Mathematical Mode of Thought in Architecture Design Education: A case study. *Nexus Network Journal* 8, 1: 93-86.

About the authors

Sarah Maor received her Ph.D. in Science and Technology Education from the Department of Education in Technology and Science. She is a Lecturer at Wizo Academy of Design and Education, Haifa, Israel.

Igor M. Verner received his Ph.D. in Computer Aided Design Systems in Manufacturing from the Department of Computer Methods and Mathematical Physics of the Urals Polytechnical Institute in Sverdlovsk, Russia. He is a Senior Lecturer in the Department of Education in Technology and Science, Technion – Israel Institute of Technology.

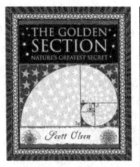

Book Review

Scott Olsen

The Golden Section:
Nature's Greatest Secret

New York: Wooden Books (Walker and Company), 2006

Reviewed by Rachel Fletcher

113 Division St.
Great Barrington, MA 01230 USA
rfletch@bcn.net

Keywords: Golden Section, phi, proportion

The Secret Life of Phi

Nothing captivates the believer or draws the skeptic's ire like geometry's Golden Section. Scott Olsen's new work on the subject is no exception.

The author's exposition on mathematics' most famous proportion is contained within a compact and beautifully assembled volume, with dozens of visual images those familiar with the topic will recognize and welcome as old friends. He begins modestly, with the simple act of dividing a line into unequal parts such that the shorter part relates to the longer in the same way that the longer relates to the whole. The result is the ratio 1:*phi* (1:ϕ or 1:$\sqrt{5}/2 + 1/2$ or 1:1.618034...), whose remarkable mathematical properties include the ability to grow at once by multiplication and simple addition.

Olsen's mission is to reveal how this simple act of division empowers individual parts to "retain a meaningful relationship to the whole," guiding spatial compositions, informing organic structures, and even illuminating the nature of consciousness. It's an ambitious agenda.

Phi plays out in pentagrams, triangles, spirals and rectangles of whirling squares. The human figure divides at the navel; fingers divide at the joints and the face into prominent features, all in Golden Section. Nature approximates the incommensurable *phi* in whole Fibonacci numbers, appearing in shells and plants, dominating phyllotactic ratios and approximating the ideal divergence angle of 137.5° ($360°/\phi^2$), conducive to packing and exposure to light. Pussy willows, sunflowers and pinecones are among countless beneficiaries.

The Golden Section would seem to offer insight into practically every living being and crafted artifact, from microscopic DNA to the macrocosm's planetary orbits. Art and architecture through the ages—Chinese cities (Beijing's Forbidden City), Egyptian temples (Karnak), Greek pottery, Gothic cathedrals (Chartres), paintings of the masters (Leonardo's *Annunciation*), and the music of Bach, Bartok and Sibelius—owe much to this proportioning technique.

Olsen draws from ancient Pythagorean, Platonic and esoteric Christian thought to show how *phi* lies at the core of human consciousness. It's a daring assertion and his spiritual

Nexus Network Journal 9 (2007) 377-378   NEXUS NETWORK JOURNAL – VOL. 9, NO. 2, 2007 **377**
1590-5896/07/020377-2 DOI 10.1007/s00004-006-0049-7
© 2007 Kim Williams Books, Turin

commentary, which must be taken on faith, will not resonate with everyone. Who can say whether the Golden Section truly unlocks great secrets in the universe? But its value in creating unified compositions—not in every situation, certainly, but where appropriate—is there for all to see.

Skeptics will argue that *phi*, being geometrically exact, leaves no room for deviation that must inevitably accompany its transfer to the material world. But one employs the Golden Section to sustain quality throughout the whole, not for quantitative exactness. Of paramount importance is whether the proportion is consistently applied. It matters little that a golden rectangle graces the facade of the Parthenon or the outline of a credit card, no matter how precise. What matters is that it divides in similar fashion through part and whole, illuminating the maker's logic and intention, while enhancing the perception of harmony; such properties are its reasons for being.

Of course, what constitute acceptable levels of tolerance is worthy of debate. Certainly, care should be taken to avoid vague or arbitrary alignments, and some studies in Olsen's book meet this criterion better than others. Critics will justifiably question his analyses of plants and animals based on fabricated drawings instead of real specimens. And some studies neglect to trace the *phi* proportion through sufficient divisions to present a convincing argument. Given the controversy that surrounds this famous proportion, a more precise and thorough citation of sources than Olsen provides is warranted.

But if specific studies are lacking, the total presentation is compelling. Olsen succeeds admirably at imparting why the proportion continues to inspire, as we marvel at the vast diversity of the living world, and how it is governed by universal principle. The book's rich and evocative design, with its splendid illustrations, deserves special mention. We've seen these images countless time before, but here they are presented as if for the very first time.

Whether the Golden Section emerges from divine inspiration, natural law, as a construct of the mind or from deep psychological need, its impact is apparent. Olsen is to be commended for presenting his case in elegant fashion. There is much here to inspire. If he tends to preach to the converted, crediting the proportion with more than it deserves, his *Golden Section* is still a worthy addition to the perennial debate.

About the reviewer

Rachel Fletcher is contributing editor to the *Nexus Network Journal* for the Geometer's Angle column.

Book Review

Leonard K. Eaton

Hardy Cross. American Engineer

Champaign, IL: University of Illinois Press, 2006

Reviewed by Kim Williams

Nexus Network Journal
Via Cavour, 8
10123 Turin (Torino) ITALY
kwilliams@kimwilliamsbooks.com

Keywords: Hardy Cross, moment distribution method, structural mechanics

It is a great pleasure for me to review this book, because I was present at its birth and watched it mature. Prof. Eaton's aim, first with an article published in the *NNJ* and then in detail with this fine new book, was to show how American engineer Hardy Cross developed a method for analyzing indeterminate structures that minimized the inconveniences and risks involved in the use and development of reinforced concrete. Along the way, however, Eaton also provides us with a tapestry of other considerations as to the interactions of engineering and architecture, the relationship of engineering and mathematics, and the contrasts between American engineers and their peers overseas. This is as culturally rich a book on a single technical argument as you could wish to find.

Leonard Eaton listened to engineer Holger Falter present a paper entitled "The Influence of Mathematics on the Development of Structural Form" at the Nexus conference on architecture and mathematics which I directed in Mantua in 1998 [Falter 1998], and took exception to one point that Falter made. In order to respond to this statement of Falter's, Prof. Eaton published a paper on the engineering achievements of Hardy Cross in the *Nexus Network Journal* [Eaton 2001]. Research for that paper led to this book.

The subject dealt with here is the problem of analyzing a statically indeterminate system. A structural system is said to be statically determinate when each span can be considered in isolation: an example is a simple beam is supported by a column at each of its ends. An indeterminate system is created by stretching the beam over three or more columns: the forces within each individual span are influenced by those of the spans adjacent to it.[1] The increasing use of reinforced concrete at the end of the nineteenth and the beginning of the twentieth century brought the problem of the analysis of indeterminate structures to the fore: construction in steel favours discontinuous elements, but construction in concrete favours the continuous.

According to engineer Falter, up through the nineteenth century the preferred methods of analyzing structure were graphic, but by that century's end analytic methods began to eclipse the graphic methods. Falter wrote, "Forming hinges was simple with iron, but difficult with reinforced concrete. It created statically indeterminate frames continuing through several spans, a difficulty surmountable only by using analytical methods" [Falter 1998: 61]. But the analysis itself could be nightmarish: "For a six-story building four bays wide, there will be six unknown x values and thirty unknown y values, or a total of thirty-six simultaneous equations to solve" [Eaton 2006: 19]. Falter's conclusion was that the difficulties inherent in the analytical methods were so enormous that engineers designed mostly simple structures, so as to keep required calculation to a minimum. Prof. Eaton countered this statement with the example of Hardy Cross and his "Moment Distribution Method", which was at least partly graphically based, and much simplified the necessary calculations.

Hardy Cross undertook to solve the problem in the early years of the 1920s. His guiding principles were beautifully simple:

- The visualization of the results of forces acting on a structure;
- The application of the theory of elasticity to structural design;
- An acceptance of approximation in lieu of completely accurate results due to inaccurate data.

Exactly how Cross addressed and solved the problems is fully explained in the book, so there is no need for me to go into that here. What I would like to do, however, is point out some of the very interesting aspects about architecture, engineering and mathematics that Prof. Eaton points out along the way.

It is well known that architects and engineers view each other with suspicion from opposite sides of the same object: that is, both professionals deal with buildings, but from points of view that are felt to be antithetical. Yet formal considerations and developments have historically gone hand-in-hand with technical considerations and developments. There are many examples of this: the great Gothic cathedrals, masterpieces both formally and structurally, is outstanding. The development of analyzing indeterminate structures to permit the use of reinforced concrete is another. Architects in the early years of the twentieth century were intrigued by the formal possibilities of reinforced concrete, but early attempts at large-scale structures were far from elegant. Prof. Eaton writes,

> ...engineering requirements seemed to dictate that increasing loads would be met by increasing the size of the structural elements. In the period from 1910 to 1930, wrote Carl Condit, "the result, which was on an elephantine scale, especially in buildings with large, open interiors, revealed the diversity of requirements that could be satisfied by reinforced concrete but also the amount of valuable space that was consumed by the sheer quantity of material in long beams and trusses." ... With his usual acuity, Hardy Cross noted that concrete was much heavier in comparison with its strength than timber or steel. "Its weight," he said, "eats up its strength in long span construction." Furthermore, it looked clumsy unless designed with skill.

How clumsy it looked will be recalled by Galileo's famous example of a bone enlarged in scale, demonstrating the "square-cube law":

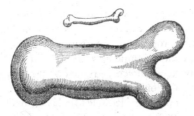

Galileo had undertaken a comparison of the strength of different-sized beams of uniform material and section. In the absence of other loads, the strength of the beam will increase with the second power, but weight of the beam will increase with the third power of the linear dimensions, that is, as the beam becomes larger, it becomes proportionally weaker, so that for strength to remain constant, it must necessary grow in size [Huerta 2006: 31].

But *how much* a reinforced concrete beam had to grow in size was unknown. And here Prof. Eaton points out what I thought was one of the most interesting cultural aspects of the problem: American engineers at the time didn't really care about finding the most elegant or efficient solution. American engineers, like their Roman counterparts two millennium earlier, merely threw more material at the problem. Eaton quotes mathematician Felix Klein, visiting the United States in about 1910, as saying,

> The United States is such a rich country that it can afford to use twice as much material in a structure as is necessary. Why should a U.S engineer be interested in a theory that tells him how to calculate exactly the amount of steel in a beam needed to carry a given load? If he's worried, all he needs to do is increase the amount of steel and make the beam stronger" [Eaton 2006: viii].

Hardy Cross apparently took up the problem because he found it intellectually intriguing; this in itself was enough to distinguish him from his compatriots.

If we are accustomed to the reciprocal diffidence of the architects and engineers, we are not so used to thinking of a diffidence on the part of engineers towards mathematics. This is the other cultural aspect of the problem that I found particularly interesting. Where theoretical mathematics was strong, so was engineering mathematics. Americans have always generally mistrusted intellectuals. Theoretical mathematics was not particularly strong in the United States, as it was more generally in Europe, and the lack of facility of engineering mathematics in the US consequently reflected that. Eaton writes that "[m]any practicing engineers had dubious mathematical skills in handling simultaneous equations and many had difficulties in visualizing rotations and displacements" [Eaton 2006: 38]. In this too Hardy Cross was distinguished from his compatriots. Cross held geometry to be of the utmost importance for the engineer; to him, the problem of indeterminate structures was geometrical. But it is a surprise to learn that Cross himself remained skeptical about the dependence of structural engineering on complicated numerical analysis. When he comments on Stefan Timoschenko's *Theory of Structures*, Cross says, "another point of view – distinctly European"; for Eaton this means "overly mathematical" [Eaton 2006, 81]. The merit of Cross's solution to the problem lies in its reduction of the necessary calculations to an easily-followed process of "solving a series of normal simultaneous equations by successive approximation" [Eaton 2006, 39].

The use of Cross's easily-applied solution to the problem of indeterminate structure opened the door for the widespread use and development of structures in reinforced concrete in American (and European) architecture. Determining the exact size that beams needed to be, eliminating useless redundancy, resulted in more elegant structures, and this surely provided an incentive to architects intrigued by the new material but concerned with aesthetics. Cross's development of the moment distribution method in the 1920s meant that when Philip Johnson and Henry-Russell Hitchcock made the International Style the new standard for architecture with their exhibit at the Museum of Modern Art in New York in 1932, American engineers were ready. With the passing years, even in the wealthy United States the economy of materials was increasingly appreciated, as during the years of the Korean War when steel was scarce. I agree with Prof. Eaton that "[i]n the field of structural engineering, Hardy Cross was the outstanding American of his time" [Eaton 2006, vii].

Hardy Cross: American Engineer is organized into a Preface, three chronologically ordered chapters dealing with Cross's career, a glossary of technical terms used, a bibliography of the works of Cross, a technical appendix by engineer Emory Kemp with worked examples of Cross's structural theories, notes, and an index. I wish there had been more illustrations. Eaton tells us that Hardy Cross thought that geometry was essential for the structural engineer, that Cross himself was forever making little drawings, and that the visualization of the results of forces on a structure was key to his clear understanding of the problem. Being able to see more of Cross's sketches would have been revealing and helpful.

Notes

1. Mario Salvadori, my mentor in many ways, said more simply that a statically determinate system is one that can be analyzed without having to appeal to any other principle than those of statics, which is the study of equilibrium, while a system that cannot be analyzed by statics alone is said to be statically indeterminate [Salvadori 1971: 4].

Bibliography

EATON, Leonard K. 2001. Hardy Cross and the "Moment Distribution Method". *Nexus Network Journal* 3, 2 (Summer-Autumn 2001): 15-24.

EATON, Leonard K. 2006. *Hardy Cross: American Engineer.* Urbana and Chicago: University of Illinois Press.

FALTER, Holger. 1998. The Influence of Mathematics on the Development of Structural Form. Pp. 51-64 in *Nexus II: Architecture and Mathematics*, Kim Williams, ed. Fucecchio, Florence: Edizioni dell'Erba.

HUERTA, Santiago. 2006. Galileo Was Wrong! The Geometrical Design of Masonry Arches. *Nexus Network Journal* 8, 2: 25-52.

SALVADORI, Mario. 1971. *Statics and Strength of Structures.* Englewood Cliffs, NJ: Prentice-Hall.

About the reviewer

Kim Williams is the editor-in-chief of the *Nexus Network Journal.*